Explore new ideas!

Welcome to the
Reading/Writing
Workshop

Read and reread exciting literature and informational texts!

Become an expert writer!

Use what you have learned to unlock the Wonders of reading!

Go Digital! www.connected.mcgraw-hill.com
Explore your Interactive Reading/Writing Workshop.

McGraw-Hill Reading
Wonders

Education

Bothell, WA • Chicago, IL • Columbus, OH • New York, NY

Cover and Title Pages: Nathan Love

www.mheonline.com/readingwonders

C

The *McGraw·Hill* Companies

 **Education**

Copyright © 2014 The McGraw-Hill Companies, Inc.

Send all inquiries to:
McGraw-Hill Education
Two Penn Plaza
New York, NY 10121

ISBN: 978-0-02-119056-0
MHID: 0-02-119056-9

Printed in the United States of America.

11 12 13 QVS 17 16 15

McGraw-Hill Reading Wonders

A Reading/Language Arts Program

Program Authors

Diane August

Donald R. Bear

Janice A. Dole

Jana Echevarria

Douglas Fisher

David Francis

Vicki Gibson

Jan Hasbrouck

Margaret Kilgo

Jay McTighe

Scott G. Paris

Timothy Shanahan

Josefina V. Tinajero

McGraw Hill Education

Bothell, WA • Chicago, IL • Columbus, OH • New York, NY

Unit 1

Think It Through

The Big Idea

How can a challenge bring out our best?........**16**

Week 1 · Clever Ideas   18

Vocabulary 20

Shared Read **The Dragon Problem** 22

Comprehension Strategy: Make Predictions.... 26

Comprehension Skill: Sequence................ 27

Genre: Fairy Tale............................... 28

Vocabulary Strategy: Synonyms................ 29

Writing: Ideas 30

Week 2 · Think of Others   32

Vocabulary 34

Shared Read **The Talent Show** 36

Comprehension Strategy: Make Predictions...... 40

Comprehension Skill: Problem and Solution...... 41

Genre: Realistic Fiction........................ 42

Vocabulary Strategy: Idioms 43

Writing: Ideas 44

(t) Alessandra Cimatoribus; (c) Valerie Sokolova; (b) Chris Vallo

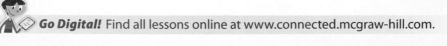

Go Digital! Find all lessons online at www.connected.mcgraw-hill.com.

Week 3 · Take Action 46

Vocabulary 48

Shared Read A World of Change 50

Comprehension Strategy: Reread 54

Comprehension Skill: Compare and Contrast 55

Genre: Expository Text 56

Vocabulary Strategy: Multiple-Meaning Words 57

Writing: Ideas 58

Week 4 · Ideas in Motion 60

Vocabulary 62

Shared Read The Big Race 64

Comprehension Strategy: Reread 68

Comprehension Skill: Cause and Effect 69

Genre: Narrative Nonfiction 70

Vocabulary Strategy: Context Clues 71

Writing: Organization 72

Week 5 · Putting Ideas to Work 74

Vocabulary 76

Shared Read TIME FOR KIDS Dollars and Sense 78

Comprehension Strategy: Reread 82

Comprehension Skill: Main Idea and Key Details 83

Genre: Persuasive Article 84

Vocabulary Strategy: Suffixes 85

Writing: Sentence Fluency 86

Unit 2 Amazing Animals

The Big Idea

What can animals teach us? **88**

Week 1 · Literary Lessons   90

Vocabulary 92

Shared Read **The Fisherman and the Kaha Bird** 94

Comprehension Strategy: Ask and Answer Questions .. 98
Comprehension Skill: Theme......................... 99
Genre: Folktale 100
Vocabulary Strategy: Root Words 101
Writing: Organization.............................. 102

Week 2 · Animals in Fiction   104

Vocabulary 106

Shared Read **The Ant and the Grasshopper** ...108

Comprehension Strategy: Ask and Answer Questions.. 112
Comprehension Skill: Theme......................... 113
Genre: Drama 114
Vocabulary Strategy: Antonyms..................... 115
Writing: Voice 116

(t) Pablo Bernasconi; (c) Amanda Hall; (b) Emily Carew Woodard

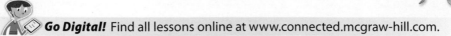

Go Digital! Find all lessons online at www.connected.mcgraw-hill.com.

Week 3 · Natural Connections 118

Vocabulary 120

Shared Read Rescuing Our Reefs 122

Comprehension Strategy: Summarize 126

Comprehension Skill: Main Idea and Key Details . . . 127

Genre: Narrative Nonfiction 128

Vocabulary Strategy: Context Clues 129

Writing: Ideas 130

Week 4 · Adaptations 132

Vocabulary 134

Shared Read Animal Adaptations 136

Comprehension Strategy: Summarize 140

Comprehension Skill: Main Idea and Key Details . . . 141

Genre: Expository Text 142

Vocabulary Strategy: Prefixes 143

Writing: Organization 144

Week 5 · Animals All Around 146

Vocabulary 148

Poetry Shared Read Dog 150

Genre: Lyric Poetry and Haiku 154

Comprehension Skill: Point of View 155

Literary Elements: Meter and Rhyme 156

Vocabulary Strategy: Similes and Metaphors 157

Writing: Word Choice 158

(t) Stephen Frink/Corbis; (c) James H. Robinson/Photolibrary; (b) Alessandra Cimatoribus

Unit 3

That's the Spirit!

The Big Idea

How can you show your community spirit? **160**

Week 1 · Friendship   162

Vocabulary 164

Shared Read **At the Library** 166

Comprehension Strategy: Visualize 170

Comprehension Skill: Point of View. 171

Genre: Fantasy 172

Vocabulary Strategy: Context Clues 173

Writing: Sentence Fluency 174

Week 2 · Helping the Community   176

Vocabulary 178

Shared Read **Remembering Hurricane Katrina** 180

Comprehension Strategy: Visualize 184

Comprehension Skill: Point of View. 185

Genre: Realistic Fiction. 186

Vocabulary Strategy: Context Clues 187

Writing: Word Choice 188

(c) Richard Johnson; (b) Jeff Mangiat

Go Digital! Find all lessons online at www.connected.mcgraw-hill.com.

Week 3 · Liberty and Justice 190

Vocabulary 192

Shared Read Judy's Appalachia 194

Comprehension Strategy: Reread 198
Comprehension Skill: Author's Point of View 199
Genre: Biography 200
Vocabulary Strategy: Synonyms and Antonyms 201
Writing: Ideas 202

Week 4 · Powerful Words 204

Vocabulary 206

Shared Read Words for Change 208

Comprehension Strategy: Reread 212
Comprehension Skill: Author's Point of View 213
Genre: Biography 214
Vocabulary Strategy: Latin and Greek Suffixes 215
Writing: Organization 216

Week 5 · Feeding the World 218

Vocabulary 220

Shared Read TIME For Kids Food Fight 222

Comprehension Strategy: Reread 226
Comprehension Skill: Author's Point of View 227
Genre: Persuasive Article 228
Vocabulary Strategy: Greek Roots 229
Writing: Voice 230

Unit 4

FACT OR FICTION?

The Big Idea

How do different writers treat the same topic?. . .**232**

Week 1 · Our Government 234

Vocabulary 236

Shared Read **A World Without Rules** 238

Comprehension Strategy: Ask and Answer Questions . . 242

Comprehension Skill: Cause and Effect 243

Genre: Narrative Nonfiction 244

Vocabulary Strategy: Latin Roots 245

Writing: Organization 246

Week 2 · Leadership 248

Vocabulary 250

Shared Read **The TimeSpecs 3000** 252

Comprehension Strategy: Make Predictions 256

Comprehension Skill: Point of View 257

Genre: Fantasy 258

Vocabulary Strategy: Idioms 259

Writing: Ideas 260

Week 3 · Breakthroughs — 262

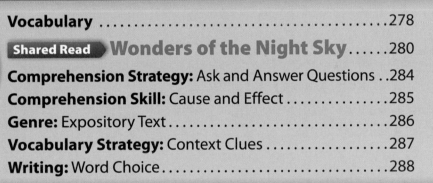

Vocabulary 264

Shared Read A Telephone Mix-Up 266

Comprehension Strategy: Make Predictions 270

Comprehension Skill: Point of View 271

Genre: Historical Fiction 272

Vocabulary Strategy: Synonyms 273

Writing: Ideas 274

Week 4 · Wonders in the Sky — 276

Vocabulary 278

Shared Read Wonders of the Night Sky 280

Comprehension Strategy: Ask and Answer Questions ..284

Comprehension Skill: Cause and Effect 285

Genre: Expository Text 286

Vocabulary Strategy: Context Clues 287

Writing: Word Choice 288

Week 5 · Achievements — 290

Vocabulary 292

Poetry **Shared Read** Sing to Me 294

Genre: Narrative Poetry 298

Comprehension Skill: Theme 299

Literary Elements: Stanza and Repetition 300

Vocabulary Strategy: Connotation and
Denotation 301

Writing: Word Choice 302

Unit 5

Figure It Out

The Big Idea

What helps you understand
the world around you?..........................304

Week 1 · Making it Happen   306

Vocabulary 308

Shared Read **Sadie's Game** 310

Comprehension Strategy: Visualize 314

Comprehension Skill: Problem and Solution....... 315

Genre: Realistic Fiction............................ 316

Vocabulary Strategy: Similes and Metaphors 317

Writing: Organization.............................. 318

Week 2 · On the Move   320

Vocabulary 322

Shared Read **My Big Brother, Johnny Kaw**..............324

Comprehension Strategy: Visualize 328

Comprehension Skill: Cause and Effect 329

Genre: Tall Tale 330

Vocabulary Strategy: Homographs.........331

Writing: Sentence Fluency 332

(t) Shelly Hehenberger; (c) James Bernardin; (b) Josée Bisaillon

Go Digital! Find all lessons online at www.connected.mcgraw-hill.com.

Week 3 · Inventions 334

Vocabulary 336

Shared Read Stephanie Kwolek:
Inventor 338

Comprehension Strategy: Summarize 342
Comprehension Skill: Problem and Solution 343
Genre: Biography 344
Vocabulary Strategy: Greek Roots 345
Writing: Sentence Fluency 346

Week 4 · Zoom In 348

Vocabulary 350

Shared Read Your World Up Close 352

Comprehension Strategy: Summarize 356
Comprehension Skill: Sequence 357
Genre: Expository Text 358
Vocabulary Strategy: Antonyms 359
Writing: Voice 360

Week 5 · Digging Up the Past 362

Vocabulary 364

Shared Read TIME FOR KIDS Where It
All Began 366

Comprehension Strategy: Summarize 370
Comprehension Skill: Sequence 371
Genre: Informational Article 372
Vocabulary Strategy: Proverbs and Adages 373
Writing: Organization 374

Unit 6 · Past, Present, and Future

The Big Idea
How can you build on what came before? **376**

Week 1 · Old and New   378

Vocabulary 380
Shared Read A Surprise Reunion 382
Comprehension Strategy: Reread 386
Comprehension Skill: Theme. 387
Genre: Historical Fiction 388
Vocabulary Strategy: Connotation and
                          Denotation 389
Writing: Word Choice. 390

Week 2 · Notes from the Past   392

Vocabulary 394
Shared Read Freedom at Fort Mose. 396
Comprehension Strategy: Reread 400
Comprehension Skill: Theme. 401
Genre: Historical Fiction 402
Vocabulary Strategy: Homophones 403
Writing: Organization. 404

(c) David McCall Johnston; (b) Neil Shigley

Go Digital! Find all lessons online at www.connected.mcgraw-hill.com.

Week 3 · Resources 406

Vocabulary 408

Shared Read The Great Energy Debate 410

Comprehension Strategy: Ask and Answer Questions.. 414

Comprehension Skill: Main Idea and Key Details 415

Genre: Narrative Nonfiction 416

Vocabulary Strategy: Latin and Greek Prefixes 417

Writing: Word Choice 418

Week 4 · Money Matters 420

Vocabulary 422

Shared Read The History of Money 424

Comprehension Strategy: Ask and Answer Questions ... 428

Comprehension Skill: Main Idea and Key Details 429

Genre: Expository Text 430

Vocabulary Strategy: Proverbs and Adages 431

Writing: Word Choice 432

Week 5 · Finding My Place 434

Vocabulary 436

Poetry Shared Read Climbing Blue Hill 438

Genre: Free Verse 442

Comprehension Skill: Theme 443

Literary Elements: Imagery and Personification 444

Vocabulary Strategy: Metaphors 445

Writing: Ideas 446

Grammar Handbook 448

(t) Reggie Casagrande/Photographer's Choice RF/Getty Images; (c) Zev Radovan/www.BibleLandPictures.com/Alamy; (b) Susan Gal

Think It Through

The BIG Idea

How can a challenge bring out our best?

The Crow and the Pitcher

A thirsty crow was flying high above the hot, dry desert when she spied a broken pitcher. The top was sharp and jagged, but she saw that there was water in the bottom. She poked her beak into the pitcher but could not reach the water so she picked up a pebble and threw it at the pitcher hoping to smash it. The pebble fell into the pitcher, hitting the water with a plop.

The angry crow picked up another pebble and was about to throw it when she stopped. She flew to the pitcher and dropped the pebble inside.

Plop!

She dropped another pebble, and another, and another.

Plop! Plop! Plop!

With each pebble the water rose closer to the top. Soon the crow was able to drink the water.

18

Creative Thinking

People come up with creative and original ideas every day. Sometimes a clever idea is the result of an accident, brainstorming, or observation.

► What do you think gave the boy in this photo the idea to build a motorcycle?

► What are some examples of clever ideas?

► Where do you get your ideas from?

Talk About It
COLLABORATE

Write words that describe how people think up ideas. Then talk to a partner about what helps you come up with good ideas.

Ideas

Vocabulary

Use the picture and the sentences to talk with a partner about each word.

brainstorm

The boys began to **brainstorm** ideas for their project.

Describe a time you had to brainstorm some ideas.

flattened

Jess enjoyed rolling out the **flattened** dough.

What is something else that can be flattened?

frantically

The dog was **frantically** digging up sand.

Describe a time when you frantically searched for something.

gracious

Justin's mom is **gracious** and kind when his friend comes over.

What is an antonym for gracious?

muttered

Dan **muttered** to himself as he read my paper.

When might you mutter something instead of saying it loudly?

official

Signing the contract will make the sale **official**.

What is an example of an official document?

original

Maria's artwork was unique and **original**.

What do you think makes something original?

stale

Grandfather and Mia threw the hard, **stale** bread out for the birds to eat.

What other kinds of food get stale?

COLLABORATE

Your Turn

Pick three words. Write three questions for your partner to answer.

Go Digital! *Use the online visual glossary*

The Dragon Problem

Essential Question

Where do good ideas come from?

Read how Liang gets a good idea.

22

Once upon a time, long before computers, baseball, or pizza, there lived a young man named Liang. During the day, Liang helped his father build furniture. At night, he made unique, **original** toys for the children in the village. He made birds with flapping wings. He carved dragons with rippling, moving scales, sharp claws, and red eyes. Every child in the village had one of Liang's dragons.

Liang knew a lot about dragons because one lived nearby on a mountain. A few times a year, the dragon would swoop down on the village. He ate water buffalo, pigs, and any people unlucky enough to be around. The Emperor had done nothing to get rid of the dragon even though his summer palace was near Liang's village.

One day in May, the Emperor and his family arrived to take up residence at his summer palace. As the procession passed through the village, the **gracious** Princess Peng smiled kindly at Liang. He fell instantly in love.

At dinner that night, Liang told his father that he wanted to marry Princess Peng. His father almost choked on the **stale**, hard rice ball he was eating.

"You're joking," his father said when he finally could speak.

"I'm serious!" insisted Liang.

His father began laughing so hard that the old chair he was sitting on broke. He lay on top of the **flattened** chair still laughing.

Valerie Sokolova

"I'll show him," Liang **muttered** angrily as he stomped out of the room.

The next morning, the Emperor's messenger made an **official** announcement.

"His Most Noble Emperor proclaims that whoever gets rid of the dragon will marry his daughter, Princess Peng."

When he heard the announcement, Liang raced to the palace to be the first to sign up. Then he looked for his friend Lee to help him **brainstorm** ideas for getting rid of the dragon. Unfortunately, Lee was away. Liang sat on a bench frowning. Nearby, children were playing with the toy dragons he had made them.

"Liang, what's wrong?" the children asked.

"I have to get rid of the dragon on the mountain," he told them.

"I have an idea," said little Ling Ling. "Why don't you carve a giant dragon and leave it by the cave? It will alarm the real dragon and scare him into flying away."

Liang stared at her. "Perfect!" he shouted and rushed home. He worked **frantically** for days making a huge, scary dragon's head. The night he finished, he loaded it onto a cart and went up the mountain. When he got near the cave, Liang put the wooden head on top of a big rock. From the front, it looked like the rest of the dragon's body was behind the rock.

Liang hid in the bushes and gave a loud roar. "What's that noise?" growled the dragon rushing out of his cave. Then he saw the massive dragon head glaring at him. "Go away, or I'll eat you up," he commanded.

The huge dragon continued to glare at him. "He must be very strong. He's not afraid of me," thought the dragon, who, like all bullies, was a coward. He decided that now was a good time to take a long trip.

"Actually, I'm leaving now. Please make yourself at home in my cave," the dragon called out as he flew away.

A year later, Liang and Princess Peng were married. They opened a toy shop together and lived happily ever after.

Make Connections

? Talk about where Liang's idea for scaring the dragon came from. **ESSENTIAL QUESTION**

Tell about a time when a friend helped you think of a good idea. **TEXT TO SELF**

Make Predictions

When you read the story "The Dragon Problem," you can use text clues and illustrations to predict what will happen next.

Find Text Evidence

As I read, I see that Liang wants to marry Princess Peng. Then the Emperor announces that anyone who gets rid of the dragon will marry his daughter. My prediction that Liang will try to get rid of the dragon was correct.

page 24

"I'll show him," Liang **muttered** angrily as he stomped out of the room.

The next morning, the Emperor's messenger made an **official** announcement.

"His Most Noble Emperor proclaims that whoever gets rid of the dragon will marry his daughter, Princess Peng."

When he heard the announcement, Liang raced to the palace to be the first to sign up. Then he looked for his friend Lee to help him **brainstorm** ideas for getting rid of the dragon. Unfortunately, Lee was away. Liang sat on a bench frowning. Nearby, children were playing with the toy dragons he had made them.

"Liang, what's wrong?" the children asked.

I read that Liang is going to the palace to sign up to get rid of the dragon. My prediction was correct.

Your Turn COLLABORATE

Make a prediction about whether the dragon will ever return to his cave. Tell what clues in the text led to your prediction. As you read, remember to use the strategy Make Predictions.

Sequence

Sequence is the order in which the key **story events** take place. Putting a story's events in sequence will help you to understand the **setting,** the **characters,** and the **plot.**

 Find Text Evidence

When I reread pages 23 and 24 of "The Dragon Problem," I see that Liang wants to marry Princess Peng. The next day, the Emperor's messenger announces that anyone who gets rid of the dragon will marry the princess.

Character
Liang

Setting
village in ancient China

Beginning
Liang *sees* Princess Peng and falls in love. The next day, the Emperor says anyone who gets rid of the dragon will marry the princess.

Middle

End

Put key story events in order to help you summarize the plot.

COLLABORATE

Your Turn

Reread "The Dragon Problem." Find the important events in the middle and end of the story. List them in the graphic organizer.

Go Digital!
Use the interactive graphic organizer

Fairy Tale

"The Dragon Problem" is a fairy tale.

Fairy tales:

- Have a main character who must complete a difficult task or journey.
- Usually contain imaginary creatures.
- Include illustrations and have a happy ending.

 Find Text Evidence

"The Dragon Problem" is a fairy tale. The story's main character must complete a difficult task. The story includes an imaginary creature, a dragon.

page 24

"I'll show him," Liang **muttered** angrily as he stomped out of the room.

The next morning, the Emperor's messenger made an **official** announcement.

"His Most Noble Emperor proclaims that whoever gets rid of the dragon will marry his daughter, Princess Peng."

When he heard the announcement, Liang raced to the palace to be the first to sign up. Then he looked for his friend Lee to help him **brainstorm** ideas for getting rid of the dragon. Unfortunately, Lee was away. Liang sat on a bench frowning. Nearby, children were playing with the toy dragons he had made them.

"Liang, what's wrong?" the children asked.

"I have to get rid of the dragon on the mountain," he told them.

"I have an idea," said little Ling Ling. "Why don't you carve a giant dragon and leave it by the cave? It will alarm the real dragon and scare him into flying away."

Liang stared at her. "Perfect!" he shouted and rushed home. He worked **frantically** for days making a huge, scary dragon's head. The night he finished, he loaded it onto a cart and went up the mountain. When he got near the cave, Liang put the wooden head on top of a big rock. From the front, it looked like the rest of the dragon's body was behind the rock.

24

Use Illustrations Fairy tales are usually illustrated. Illustrations give visual clues about the characters, settings, and events in the story.

COLLABORATE

Your Turn

With a partner, discuss whether the ending is surprising for a fairy tale. Explain why or why not.

Synonyms

As you read "The Dragon Problem," you may come across a word that you don't know. Look at the surrounding words and sentences for clues. Sometimes the author uses a synonym, a word that means almost the same thing as the unfamiliar word.

 Find Text Evidence

When I read the third sentence on page 23 in "The Dragon Problem," the word original *helps me to figure out what the word* unique *means.*

At night, he made unique, original toys for the children in the village.

Your Turn

COLLABORATE

Look for synonyms to find the meanings of the following words in "The Dragon Problem."

rippling, *page 23*
alarm, *page 24*
massive, *page 25*

Readers to . . .

Writers include specific, concrete, and sensory details when writing stories. These details provide a visual picture for the reader. Reread the excerpt below from "The Dragon Problem."

Expert Model

Descriptive Details

Identify the **descriptive details** in the story. How do the details help readers picture what is happening in the story?

During the day, Liang helped his father build furniture. At night, he made unique, original toys for the children in the village. He made birds with flapping wings. He carved dragons with rippling, moving scales, sharp claws, and red eyes. Every child in the village had one of Liang's dragons.

Writers

Martin wrote a story about a prince. Read Martin's revisions to a section of his story.

Student Model

THE LOST PRINCE

Once ^(in a faraway kingdom) there was a prince who always got lost. He turned left when he meant to turn right. He walked ten miles instead of one. He got lost a lot. ^(every day of his life)

The King hired a wise man to help his son. First, the wise man ^wrote an R on the prince's right hand and an L on his left. Next, he gave him a compass ^(that showed all four directions). Finally, he invented a machine to tell the prince when he had gone too far.

Editing Marks

⌐┘ Switch order.

∧ Add.

⌃ Add a comma.

⊙ Add a period.

⌇ Take out.

(SP) Check spelling.

≡ Make a capital letter.

Grammar Handbook

Sentences
See page 450.

Your Turn

COLLABORATE

✔ Identify the details.
✔ Did Martin use complete sentences?
✔ Tell how revisions improved his writing.

Go Digital!
Write online in Writer's Workspace

Chris Vallo

Essential Question

How do your actions affect others?

Go Digital!

Actions Count

Have you ever heard the expression, "Actions speak louder than words"? A broken promise is one example of actions speaking louder than words. Can you name another example?

▶ How would you feel if you were sitting next to these two girls?

▶ When have your actions affected friends or family in either a good way or a bad way?

Talk About It

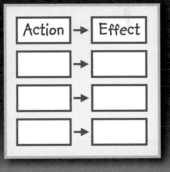

COLLABORATE

List some of your actions and the effects they have had on the people around you. Then talk with a partner about how your actions can affect others.

Action	→	Effect
	→	
	→	
	→	

Vocabulary

Use the picture and the sentences to talk with a
partner about each word.

accountable

Sam is held **accountable** for washing
his dog.

How are the words accountable and
responsible similar?

advise

A coach can **advise** you on how to
improve your swimming.

What is a synonym for advise?

desperately

The woman was **desperately** trying to
remember where she had left her keys.

Describe a time when you desperately
tried to remember something.

hesitated

The dog **hesitated** before jumping up
to grab the food off the counter.

When have you hesitated before doing
something?

humiliated

Sarah felt **humiliated** when she forgot her lines.

How is humiliated similar to embarrassed?

inspiration

The girl found **inspiration** for her drawing in nature.

When you have to write a story where does your inspiration come from?

self-esteem

Winning the soccer championship helped improve Billy's confidence and **self-esteem**.

What else builds self-esteem?

uncomfortably

Sonya's throat felt **uncomfortably** sore.

What are some things that can feel uncomfortably tight?

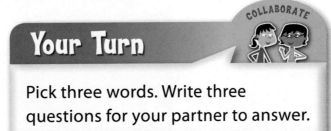

Your Turn COLLABORATE

Pick three words. Write three questions for your partner to answer.

Go Digital! *Use the online visual glossary*

THE Talent Show

? Essential Question

How do your actions affect others?

Read about how Tina's actions affect Maura.

"Tina, there's a school talent show in three weeks," I shouted to my best friend. My older brother had been teaching me juggling, and I knew he'd help me with my act for the show.

Tina ran over to the bulletin board and read the poster. "Maura, what's our act going to be?" Tina asked me.

"Our act?" I said, taking a tighter grip on my books.

Tina grinned, pointed to the poster and said "It says acts can be individuals, partners, or small groups."

My grip on my books became uncomfortably tight. "You want to do an act together?"

"It'll be fun," Tina said.

I hesitated for a second before continuing. "I've got an idea and. . . ."

Tina interrupted me. "Yeah, me too; let's talk at lunch."

During math, I tried to think of how I would tell Tina that I wanted to do my own act. After all, we are best friends; we should be able to see eye to eye about this. The problem is Tina always takes charge, I don't speak up, and then I end up feeling resentful about the whole situation.

I desperately wanted to win, but it was more than that. I wanted to win on my own—with an act that was all mine.

Chris Vallo

At lunch, Tina started talking as soon as we sat down. "I have it all planned out. My inspiration came from that new TV show, 'You've Got Talent.' We can sing along to a song and do a dance routine, and my mother can make us costumes."

"Yeah, that's good," I said. "But I had another idea." I told her about my juggling act.

Tina considered it. "Nah, I don't think I can learn to juggle in three weeks and I'd probably drop the balls," she said. "We don't want to be humiliated, right?"

At recess, I ran around the track a couple of times just to let off steam.

When my grandmother picked me up after school, she drove a few minutes and finally said, "Cat got your tongue?"

I explained about the talent show as she listened carefully. "So, Tina is not being respectful of your ideas, but it sounds as if you aren't either."

"What?" I shouted. "I told Tina her idea was good."

"No," said my grandmother, "I said that you weren't respectful of your *own* ideas, or you would have spoken up. I understand that you're friends, but you're still accountable for your own actions."

I thought about this. "So what should I do?" I asked.

"I **advise** you to tell the truth," she said. "It wouldn't hurt to let Tina know what you want. Besides," my grandmother added, "it will be good for your **self-esteem**!"

When we got home, I took 12 deep breaths, called Tina, and told her that I was going to do my juggling act. She was curt on the phone, and I spent all night worrying she would be mad at me.

The next day, she described her act and her costume. But the biggest surprise came at recess, when we played a game that I chose, not Tina.

I guess standing up for myself did pay off.

Make Connections

Talk about how Maura was affected by Tina's actions. **ESSENTIAL QUESTION**

Tell about a time when someone wouldn't listen to your ideas. What did you do? **TEXT TO SELF**

Make Predictions

When you read, use story details to make predictions about what will happen. As you read "The Talent Show," make predictions.

 Find Text Evidence

You predicted that Tina is the kind of friend who is bossy. Reread page 37 of "The Talent Show" to find the text evidence that confirms your prediction.

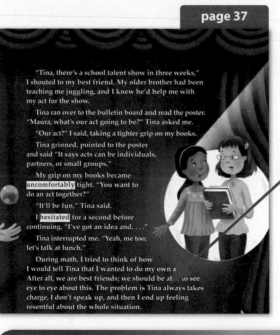

page 37

"Tina, there's a school talent show in three weeks," I shouted to my best friend. My older brother had been teaching me juggling, and I knew he'd help me with my act for the show.

Tina ran over to the bulletin board and read the poster. "Maura, what's our act going to be?" Tina asked me.

"Our act?" I said, taking a tighter grip on my books.

Tina grinned, pointed to the poster and said "It says acts can be individuals, partners, or small groups."

My grip on my books became uncomfortably tight. "You want to do an act together?"

"It'll be fun," Tina said.

I hesitated for a second before continuing. "I've got an idea and...."

Tina interrupted me. "Yeah, me too; let's talk at lunch."

During math, I tried to think of how I would tell Tina that I wanted to do my own act. After all, we are best friends; we should be able to see eye to eye about this. The problem is Tina always takes charge, I don't speak up, and then I end up feeling resentful about the whole situation.

I read that Tina always takes charge. This confirms my prediction that Tina is bossy.

Your Turn

COLLABORATE

Using clues you find in the text, how do you predict Maura will solve a future problem? As you read, use the strategy Make Predictions.

Problem and Solution

The main **character** in a story usually has a problem that needs to be solved. The steps the character takes to solve the problem make up the **story's events**, the plot of the story.

 Find Text Evidence

As I reread pages 37 and 38 of "The Talent Show," I can see that Maura has a problem. I will list the events in the story. Then I can figure out how Maura finds a solution.

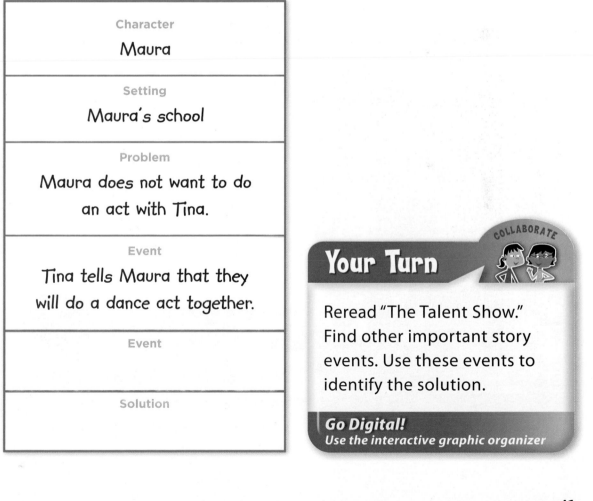

Character
Maura

Setting
Maura's school

Problem
Maura does not want to do an act with Tina.

Event
Tina tells Maura that they will do a dance act together.

Event

Solution

Your Turn COLLABORATE

Reread "The Talent Show." Find other important story events. Use these events to identify the solution.

Go Digital!
Use the interactive graphic organizer

Realistic Fiction

The selection "The Talent Show" is realistic fiction.

Realistic Fiction:
- Is a made-up story.
- Has characters, settings, and events that could exist in real life.
- Includes dialogue.

Find Text Evidence

I can tell that "The Talent Show" is realistic fiction. The story mostly takes place at school. On page 37, Maura and Tina act and speak like real people who might go to my school.

page 37

"Tina, there's a school talent show in three weeks," I shouted to my best friend. My older brother had been teaching me juggling, and I knew he'd help me with my act for the show.

Tina ran over to the bulletin board and read the poster. "Maura, what's our act going to be?" Tina asked me.

"Our act?" I said, taking a tighter grip on my books.

Tina grinned, pointed to the poster and said "It says acts can be individuals, partners, or small groups."

My grip on my books became uncomfortably tight. "You want to do an act together?"

"It'll be fun," Tina said.

I hesitated for a second before continuing. "I've got an idea and. . . ."

Tina interrupted me. "Yeah, me too; let's talk at lunch."

During math, I tried to think of how I would tell Tina that I wanted to do my own act. After all, we are best friends; we should be able to see eye to eye about this. The problem is Tina always takes charge, I don't speak up, and then I end up feeling resentful about the whole situation.

I desperately wanted to win, but it was more than that. I wanted to win on my own—with an act that was all mine.

37

Dialogue Dialogue is the exact words the characters say.

Your Turn

COLLABORATE

With a partner, list two examples from "The Talent Show" that let you know it is realistic fiction.

Idioms

Idioms are phrases that have a meaning different from the meaning of each word in them. Sometimes context clues can help you figure out the meaning of an idiom.

 Find Text Evidence

When I read the idiom see eye to eye *on page 37 in "The Talent Show," the words* After all, we are best friends *help me figure out its meaning. To* see eye to eye *means to agree.*

After all, we are best friends; we should be able to see eye to eye about this.

Your Turn

COLLABORATE

Use context clues to help you understand the meanings of the following idioms in "The Talent Show":

let off steam, *page 38*

cat got your tongue, *page 38*

standing up for myself, *page 39*

List some other idioms and their meanings.

Chris Vallo

Readers to...

Writers know that many small moments make up an event. When a writer focuses on an event, he or she describes the small moments that create the event. Reread the excerpt from "The Talent Show" below.

Expert Model

Focus on an Event

Identify the **event**. What small moments help to describe the event?

"Our act?" I said, taking a tighter grip on my books.

Tina grinned and pointed to the poster. "It says acts can be individuals, partners, or small groups."

My grip on my books became uncomfortably tight. "You want to do an act together?"

"It'll be fun," Tina said.

I hesitated for a second before continuing. "I've got an idea and . . . ."

Tina interrupted me. "Yeah, me, too! Let's talk at lunch."

Chris Vallo

44

Writers

Kyra wrote a story about two friends. Read Kyra's revisions to a section of her story.

Editing Marks

⊔ Switch order.

∧ Add.

⌄ Add a comma.

⌇ Take out.

(SP) Check spelling.

≡ Make a capital letter.

Grammar Handbook

Subjects and Predicates See page 451.

Student Model

Dogs or Kids?

    and Selena
Nan ∧ needed to earn some extra money. ~~Selena needed to earn money, too.~~

    "How about baby-sitting?" Nan asked when they were trying to come up with ideas. Selena's eyes lit up, and she smiled. "How about ∧ *a* *business* dog-walking ∧?"

    Nan frowned. "But I am allergic to dogs. That idea won't work for me."

Your Turn

COLLABORATE

☑ Identify the event.
☑ Find a compound subject.
☑ Tell how revisions improved Kyra's writing.

Go Digital!
Write online in Writer's Workspace

Chris Vallo

? Essential Question
How do people respond to
natural disasters?

Go Digital!

To the Rescue

Natural disasters are events such as hurricanes, earthquakes, floods, and forest fires. When these kinds of events occur, it can cause a huge crisis in a community. Luckily, there are people who are trained to respond to natural disasters.

▶ How might people respond to a forest fire?

▶ How do you think people are rescued during a flood?

▶ What are some ways that people might respond during other kinds of natural disasters?

Talk About It

COLLABORATE

Write words you have learned about responding to natural disasters. Then talk to a partner about what you might do to help after a natural disaster.

Natural Disasters

Vocabulary

Use the picture and the sentences to talk with a partner about each word.

alter

The ocean waves slowly **alter** the shoreline by carving away the rocks.

How can people alter their appearance?

collapse

Flood waters caused the bridge to **collapse**.

What might cause a tent to collapse?

crisis

Rescue workers help people during an emergency or a **crisis,** such as a flood.

How would you react to a crisis?

destruction

The tornado destroyed buildings and caused a lot of other **destruction**.

What is a synonym for destruction?

hazard

The water was a **hazard** to people driving on the street.

What else might be a hazard to people who are driving?

severe

Severe weather can include very strong winds and heavy rain.

Describe severe winter weather.

substantial

We got a **substantial** amount of snow last night.

What is an antonym for substantial?

unpredictable

The **unpredictable** weather turned suddenly from sun to rain.

What is an antonym for unpredictable?

COLLABORATE

Your Turn

Pick three words. Write three questions for your partner to answer.

Go Digital! *Use the online visual glossary*

A World of CHANGE

? Essential Question

How do people respond to natural disasters?

Read about how people prepare for natural disasters.

The Grand Canyon Skywalk, Arizona

Earth may seem as if it is a large rock that never changes. Actually, our planet is in a constant state of change. Natural changes take place every day. These activities **alter** the surface of Earth. Some of these changes take place slowly over many years. Others happen in just minutes. Whether they are slow or fast, both kinds of changes have a great effect on our planet.

Slow and Steady

Some of Earth's biggest changes can't be seen. That is because they are happening very slowly. Weathering, erosion, and deposition are three natural processes that change the surface of the world. They do it one grain of sand at a time.

Weathering occurs when rain, snow, sun, and wind break down rocks into smaller pieces. These tiny pieces of rock turn into soil, but they are not carried away from the landform.

Erosion occurs when weathered pieces of rock are carried away by a natural force such as a river. This causes landforms on Earth to get smaller. They may even completely **collapse** over time. The Grand Canyon is an example of the effect of erosion. It was carved over thousands of years by the Colorado River.

After the process of erosion, dirt and rocks are then dropped in a new location. This process is called deposition. Over time, a large collection of deposits may occur in one place. Deposition by water can build up a beach. Deposition by wind can create a **substantial** landform, such as a sand dune.

(bkgd) Julie Quarry/Alamy; (titles) image100/Corbis

Although erosion is a slow process, it still creates problems for people. Some types of erosion are dangerous. They can be seen as a hazard to communities.

To help protect against beach erosion, people build structures that block ocean waves from the shore. They may also use heavy rocks to keep the land from eroding. Others grow plants along the shore. The roots of the plants help hold the soil and make it less likely to erode.

Unfortunately, people cannot protect the land when fast natural processes occur.

Fast and Powerful

Fast natural processes, like slow processes, change the surface of Earth. But fast processes are much more powerful. They are often called natural disasters because of the destruction they cause. Volcanic eruptions and landslides are just two examples.

Volcanoes form around openings in Earth's crust. When pressure builds under Earth's surface, hot melted rock called magma is forced upwards. It flows up through the volcano and out through the opening. Eruptions can occur without warning. They have the potential to cause a crisis in a community.

Like volcanic eruptions, landslides can happen without warning. They occur when rocks and dirt, loosened by heavy rains, slide down a hill or mountain. Some landslides are small. Others can be quite large and cause severe damage.

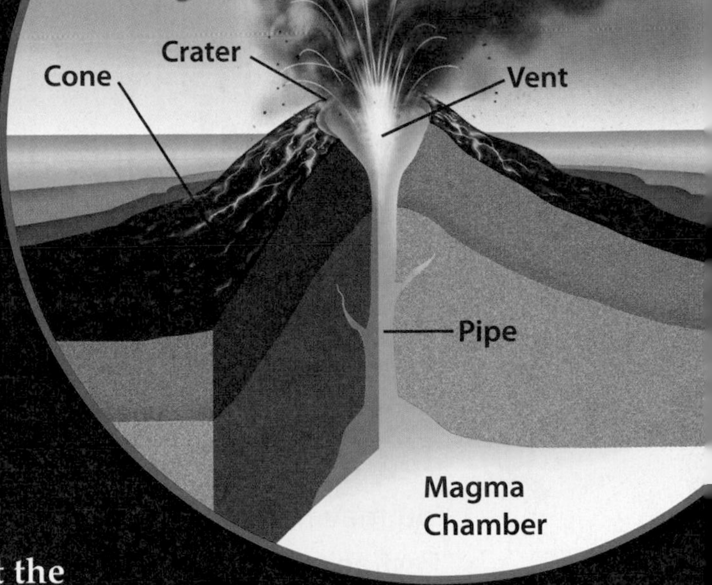

Be Prepared

In contrast to slow-moving processes, people cannot prevent the effects of fast-moving natural disasters. Instead, scientists try to predict when these events will occur so that they can warn people. Still, some disasters are unpredictable and strike without warning. It is important for communities to have an emergency plan in place so that they can be evacuated quickly.

The surface of Earth constantly changes through natural processes. These processes can be gradual or swift. They help to make Earth the amazing planet that it is!

This diagram shows a volcano erupting.

Cone
Crater
Vent
Pipe
Magma Chamber

Make Connections

Talk about different ways that people prepare for natural disasters. **ESSENTIAL QUESTION**

How can you help others who have been in a natural disaster? **TEXT TO SELF**

Reread

When you read an informational text, you may come across facts and ideas that are new to you. As you read "A World of Change," you can reread the difficult sections to make sure you understand them and to help you remember key details.

Find Text Evidence

You may not be sure why a volcano erupts. Reread the section "Fast and Powerful" on page 52 of "A World of Change."

page 52

Fast and Powerful

Fast natural processes, like slow processes, change the surface of Earth. But fast processes are much more powerful. They are often called natural disasters because of the destruction they cause. Volcanic eruptions and landslides are just two examples.

Volcanoes form around openings in Earth's crust. When pressure builds under Earth's surface, hot melted rock called magma is forced upwards. It flows up through the volcano and out through the opening. Eruptions can occur without warning. They have the potential to cause a crisis in a community.

> *I read that when pressure builds under Earth's surface, magma is forced upwards. From this I can draw the inference that pressure below the surface causes a volcano to erupt.*

COLLABORATE

Your Turn

What happens to rock during weathering? Reread the section "Slow and Steady" on page 51 to find out. As you read, remember to use the strategy Reread.

Compare and Contrast

Authors use text structure to organize the information in a text. Comparison is one kind of text structure. Authors who use this text structure show how things are alike and different.

 Find Text Evidence

Looking back at pages 51–52 of "A World of Change, " I can reread to learn how slow natural processes and fast natural processes are alike and different. Words such as some, but, both, *and* like *let me know that a comparison is being made.*

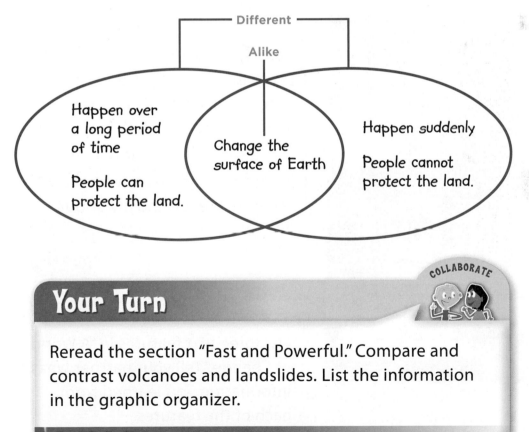

Different

Alike

Happen over a long period of time

People can protect the land.

Change the surface of Earth

Happen suddenly

People cannot protect the land.

COLLABORATE

Your Turn

Reread the section "Fast and Powerful." Compare and contrast volcanoes and landslides. List the information in the graphic organizer.

Go Digital! *Use the interactive graphic organizer*

Expository

The selection "A World of Change" is an expository text.

Expository text:
- Explains facts about a topic.
- Includes text features.

Find Text Evidence

"A World of Change" is an expository text. It gives many facts about Earth's processes. Each section has a heading that tells me what the section is about. The diagram gives me more information.

---

page 53

This diagram shows a volcano erupting.

Cone Crater Vent
Pipe
Magma Chamber

Like volcanic eruptions, landslides can happen without warning. They occur when rocks and dirt, loosened by heavy rains, slide down a hill or mountain. Some landslides are small. Others can be quite large and cause severe damage.

Be Prepared

In contrast to slow-moving processes, people cannot prevent the effects of fast-moving natural disasters. Instead, scientists try to predict when these events will occur so that they can warn people. Still, some disasters are unpredictable and strike without warning. It is important for communities to have an emergency plan in place so that they can be evacuated quickly.

The surface of Earth constantly changes through natural processes. These processes can be gradual or swift. They help to make Earth the amazing planet that it is!

Make Connections

Talk about different ways that people prepare for natural disasters. ESSENTIAL QUESTION

How can you help others who have been in a natural disaster? TEXT TO SELF

53

Text Features

Diagrams Diagrams show the parts of something or how a process works. They have labels that tell about their different parts.

Headings Headings tell what a section of text is mostly about.

COLLABORATE

Your Turn

List three text features in "A World of Change." Tell your partner what information you learned from each of the features.

Multiple-Meaning Words

As you read "A World of Change," you will come across some **multiple-meaning words**. These are words that have more than one meaning. To figure out the meaning of a multiple-meaning word, check the words and phrases near it for clues.

 Find Text Evidence

When I read page 52 of "A World of Change," I see the word block. *There are a few different meanings for* block, *so this is a multiple-meaning word. The word* protect *and the phrase "ocean waves from the shore" help me figure out which meaning is being used in the sentence.*

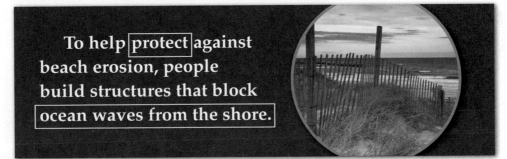

To help protect against beach erosion, people build structures that block ocean waves from the shore.

Your Turn

COLLABORATE

Use context clues to figure out the meanings of the following words in "A World of Change."

place, *page 51*
shore, *page 52*
strike, *page 53*

Readers to...

Writers make sure they focus on a topic by providing a main idea when they write expository text. They use important details to support the main idea. Reread the first paragraph of "A World of Change" below.

Focus on a Topic

Identify supporting details for the main idea that Earth is in a constant state of change.

Expert Model

A World of Change

Earth may seem as if it is a large rock that never changes. Actually, our planet is in a constant state of change. Natural changes take place every day. These activities alter the surface of Earth. Some of these changes take place slowly over many years. Others happen in just minutes. Whether they are slow or fast, both kinds of changes have a great effect on our planet.

Writers

Jake wrote an expository text. Read Jake's revision of a section of it.

Editing Marks

⌐⌐ Switch order.

∧ Add.

∧⸴ Add a comma.

ℒ Take out.

(SP) Check spelling.

≡ Make a capital letter.

Grammar Handbook

Compound Sentences See page 451.

Student Model

Yellowstone National Park

Yellowstone National Park is
Millions of people visit this park every year.
popular. People come from all over

the world.

Yellowstone is a beautiful park

to visit. People photograph the

waterfalls and the animals, and they
This famous geyser erupts every 1–2 hours.
make sure to visit Old Faithful, too. ∧

There are lots of different

animals at Yellowstone,

including elk, bison,

and grizzly bears.

Your Turn

COLLABORATE

- ☑ Identify the details that Jake included.
- ☑ Identify a compound sentence.
- ☑ Tell how Jake's revisions improved his writing.

Go Digital!
Write online in Writer's Workspace

Essential Question

How can science help you understand how things work?

Go Digital!

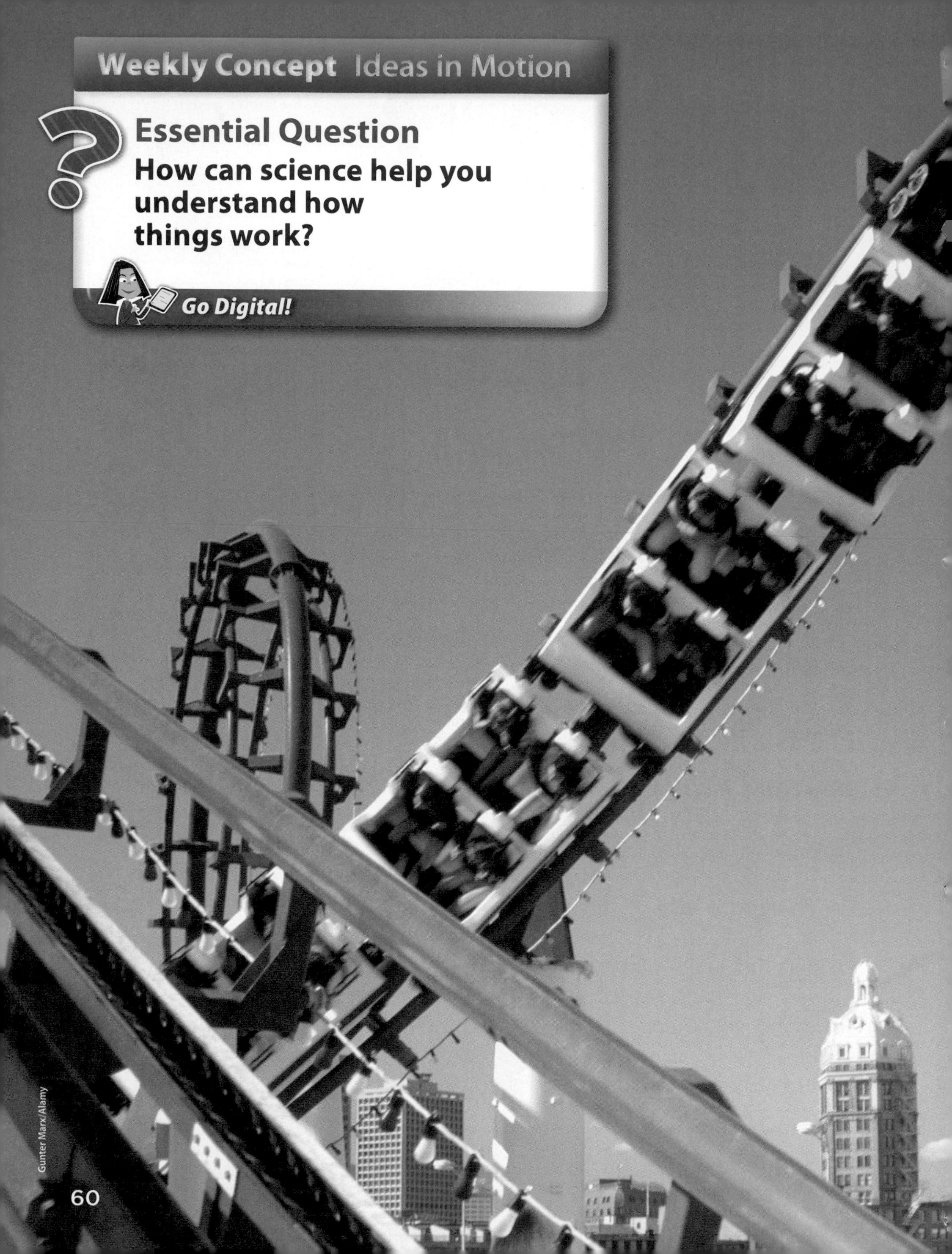

60

HOW DOES IT WORK?

Science can help us understand a lot of things—from how to throw a curve ball to what happens when you ride a roller coaster. Look at this picture. What keeps these people from falling out? Let's use science to find out!

► How do you stay in place during the loop-the-loops? The force created by the acceleration presses you against the seat of the coaster.

► What kind of rides have you ridden on at an amusement park? Why did you like them?

Talk About It

Write words that you have learned about motion. Talk to your partner about a ride that you would design.

Motion

Vocabulary

Use the picture and the sentences to talk with a partner about each word.

accelerate

I saw the race car **accelerate,** or speed up, across the finish line.

What is an antonym for accelerate?

advantage

The father's size gave him a big **advantage** over his son.

What is a synonym for advantage?

capabilities

The **capabilities** of a potter include strength and creativity.

What capabilities would an athlete need?

friction

The **friction** between the tires and the pavement slows down the airplane.

How is using the brakes on a bike an example of friction?

gravity

Gravity helps pull the batter down into the baking pan.

Describe what would happen if there were no gravity on Earth.

identity

The woman showed her passport to prove her **identity**.

Why might somebody want to keep his or her identity a secret?

inquiry

Reporters ask questions at the beginning of any **inquiry** or investigation.

How are the words inquiry and investigation similar?

thrilling

Going on a roller coaster can be exciting and **thrilling**.

What is an antonym for thrilling?

COLLABORATE

Your Turn

Pick three words. Write three questions for your partner to answer.

Go Digital! *Use the online visual glossary*

Alex and Liam planned to build a car for the soap box derby. As a result of their **inquiry** into how to build a fast car, they had come to the science museum today for answers. Last week, Alex's mother had called one of the museum's scientists. When they walked into the museum, a woman in a lab coat and inline skates zoomed up and greeted them.

"Hi, I'm Clara. Are you the boys who want to know what will make a car go fast?"

"Yes, I'm Alex, and that's Liam," Alex responded.

"Why are you wearing inline skates, Clara?" Liam asked.

"I'm a champion skater!" Clara claimed, doing a spin. Then she whispered, "That's not my true **identity**. I'm a scientist. Skates make it easier to get around. Follow me!"

IT'S ABOUT SPEED

"Welcome to our On the Move exhibit," Clara announced as they entered a large room. "So, tell me about the race."

"There will be 20 cars in the race. We'll be going down the steepest hill in town!" Alex said.

"Sounds **thrilling**! It must be exciting to go fast!" Clara answered as she pressed buttons on a machine. "This is a virtual race car, and this screen shows you the virtual race course and your speed. Speed is the distance an object moves in a certain amount of time."

Craig Phillips

FORCES AT WORK

Alex and Liam climbed into the machine. Each seat had a steering wheel and a screen in front of it.

A force is a push or pull.

Clara said, "Since you want to build a fast car, you need to know about forces and how they affect motion."

"What's a force?" asked Liam.

Clara continued, "A force is a push or a pull. Forces cause things to move or cause a change in motion. When I apply a big enough force on an object, like this stool, it moves. If two objects are exactly the same, the object that receives a bigger force will **accelerate**, or increase its speed," Clara said, pushing two stools at the same time.

"Which stool received a bigger force?" Clara asked.

"The one on the right. It went farther," said Liam.

"So, giving our car a big push at the top of the hill will cause it to accelerate and go faster," Alex summarized.

There's a sharp curve coming up!

I'm going to accelerate now!

GRAVITY AND FRICTION

Clara smiled, "Right! Another force acting on your car is **gravity**. Gravity is a pulling force between two objects." Clara took a tennis ball out of her pocket. "When I drop this ball, gravity pulls it towards the floor. It's the same force that pulls your car down the hill."

"So, a big push gives us an **advantage** over other cars, and gravity will keep us going. How do we stop?" Liam asked.

"You'll need **friction**. Friction is a force between two surfaces that slows objects down or stops them from moving. For example, I lean back on my skates, and the friction between the rubber stoppers and the floor slows me down," said Clara.

"Thanks, Clara! The virtual race car was cool! I knew we had the skills and **capabilities** to win the race, but now we have science on our side, too," Liam grinned.

You need friction.

Make Connections

Talk about ways that science can help you understand how objects move. **ESSENTIAL QUESTION**

How can science help you understand your favorite activities? **TEXT TO SELF**

Craig Phillips

67

Reread

When you read an informational text, you often come across information that is new to you. As you read "The Big Race," reread key sections of text to make sure you understand them and remember the information they contain.

Find Text Evidence

As you read "The Big Race," the concept of acceleration may be new to you. Reread the "Forces at Work" section on page 66 to help you remember what *accelerate* means.

page 66

"What's a force?" asked Liam.

Clara continued, "A force is a push or a pull. Forces cause things to move or cause a change in motion. When I apply a big enough force on an object, like this stool, it moves. If two objects are exactly the same, the object that receives a bigger force will **accelerate**, or increase its speed," Clara said, pushing two stools at the same time.

"Which stool received a bigger force?" Clara asked.

"The one on the right. It went farther," said Liam.

"So, giving our car a big push at the top of the hill will cause it to accelerate and go faster," Alex summarized.

There's a sharp curve coming up!

I'm going to accelerate now!

I read that accelerate *means to increase the speed of something. Rereading will help me to understand and remember this concept.*

Your Turn

COLLABORATE

What does gravity do? Reread the "Gravity and Friction" section of "The Big Race" to find out. As you read, remember to use the strategy Reread.

Cause and Effect

Text structure is the way that authors organize information in a selection. Cause and effect is one kind of text structure. The author explains how and why something happens. A cause is why something happens. An effect is what happens.

Find Text Evidence

I can reread "Forces at Work" in "The Big Race" on page 66 to find actions that cause something to happen. Then I can figure out the effects of those actions.

Cause	→	Effect
Clara applies force to one stool.	→	The stool moves.
Clara pushes both stools.	→	Both stools move.
Clara applies more force to one of the stools.	→	One stool moves farther.

COLLABORATE

Your Turn

Reread each section of "The Big Race." Find events or actions that cause something to happen and their effects. List each cause and effect in the graphic organizer.

Go Digital!
Use the interactive graphic organizer

Narrative Nonfiction

The selection "The Big Race" is narrative nonfiction.

Narrative nonfiction:
- Tells a story.
- Includes facts and examples about a topic.
- Often includes text features.

Find Text Evidence

Even though "The Big Race" reads like a story, I can tell that it is an informational text because it includes facts and text features.

page 66

FORCES AT WORK

Alex and Liam climbed into the machine. Each seat had a steering wheel and a screen in front of it.

Clara said, "Since you want to build a fast car, you need to know about forces and how they affect motion."

"What's a force?" asked Liam.

Clara continued, "A force is a push or a pull. Forces cause things to move or cause a change in motion. When I apply a big enough force on an object, like this stool, it moves. If two objects are exactly the same, the object that receives a bigger force will **accelerate**, or increase its speed," Clara said, pushing two stools at the same time.

"Which stool received a bigger force?" Clara asked.

"The one on the right. It went farther," said Liam.

"So, giving our car a big push at the top of the hill will cause it to accelerate and go faster," Alex summarized.

A force is a push or pull.

There's a sharp curve coming up!

I'm going to accelerate now!

66

Text Features

Headings Headings tell what a section of text is mostly about.

Speech Balloons Speech Balloons tell what the characters are saying or thinking.

COLLABORATE

Your Turn

Find two examples of text features in "The Big Race." Tell your partner what information you learned from the features.

Context Clues

When you are not sure what a word means, you can look at the other words around it to figure out the meaning. These other words, called context clues, may be **definitions,** **examples**, or **restatements** of the word's meaning.

 Find Text Evidence

When I read the fourth paragraph on page 66 of "The Big Race," I am not sure what the word force *means. The phrase "a push or a pull" defines what the word* force *means.*

Clara continued, "A force is a push or a pull. Forces cause things to move or cause a change in motion."

Your Turn

Use context clues to figure out the meanings of the following words in "The Big Race":

speed, *page 65*
friction, *page 67*
surfaces, *page 67*

Readers to...

Writers choose the best way to organize their information. One way to organize information is to present events in the order in which they happen. Reread the section from "The Big Race" below.

Sequence

Identify the **sequence** of information in this excerpt. What time-order words does the author use?

Expert Model

Alex and Liam planned to build a car for the soap box derby. As a result of their inquiry into how to build a fast car, they had come to the science museum today for answers. Last week, Alex's mother had called one of the museum's scientists. When they walked into the museum, a woman in a lab coat and inline skates zoomed up and greeted them.

Writers

Editing Marks

⊓ Switch order.

∧ Add.

∧⸝ Add a comma.

⊙ Add a period.

✐ Take out.

(SP) Check spelling.

≡ Make a capital letter.

Jonah wrote a narrative nonfiction piece. Read Jonah's revision of a section of it.

Student Model

Grammar Handbook

Complex Sentences
See page 453.

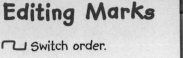

BOATING

My grandfather asked me to go boating today. I love to go boating, ∧^but^ I was scared we would sink. First, he told me about buoyancy force⊙∧

Buoyancy force pushes an object

^For example, if I drop a ball in the water it won't sink.^ upward when it is in water. ∧Buoyancy forces it up to the water's surface.

Now, I wanted to go boating!

Next, he explained that the lake was not very deep. ←

Your Turn

- ✔ How does Jonah present new information?
- ✔ Identify a complex sentence he uses.
- ✔ How did the revisions improve his writing?

Go Digital!
Write online in Writer's Workspace

? **Essential Question**
How can starting a business
help others?

Go Digital!

TIME FOR KIDS®

RISE

TO THE CHALLENGE

How do you start a business and help people at the same time? A woman in New York did it. She started a bakery that includes a culinary training program for immigrants. Not only has the training program been successful, the bakery's breads are a big hit too.

▶ How do you think a business can give back to the community? What kinds of things could they do?

▶ What kind of business would you start? How would it help people or your community?

Talk About It

Write words that tell how starting a business can help people. Then talk about a business you would like to start.

Starting a Business

Vocabulary

Use the picture and the sentences to talk with a partner about each word.

compassionate

I could tell she was a **compassionate** and caring person by the way she hugged her sister.

What is an antonym for compassionate?

enterprise

Starting a white water rafting business was an exciting new **enterprise** for Tom.

What is the first thing a person starting a new enterprise might do?

exceptional

Monica is an **exceptional** and talented flute player.

How does a person become exceptional at doing something?

funds

Nicole's class held a bake sale to raise **funds** to buy books for the library.

What project would you like to raise funds for?

innovative

Sam enjoyed trying out the new **innovative** racing wheelchair.

What new technology do you think is innovative?

process

An important step in the **process** of making a pie is to roll out the crust.

What is one step in the process of baking cookies?

routine

Brittany loved the daily **routine** of walking her dog.

Why is it helpful to have a morning routine?

undertaking

Cleaning up Tim's messy bedroom was going to be a big **undertaking**.

What would you consider a big undertaking?

COLLABORATE

Your Turn

Pick three words. Write three questions for your partner to answer.

Go Digital! **Use the online visual glossary**

Essential Question

How can starting a business help others?

Read about how two companies are making a difference.

78

Dollars and $ENSE

Behind the success of these big businesses is a desire to help others.

Good business is not always about the bottom line. A **compassionate** company knows that making money is not the only way to measure success. Many large businesses in the United States and all over the world are finding unusual ways to help people in need.

Hearts and Soles

After starting and running four businesses, Blake Mycoskie wanted a break from his usual **routine**. In 2006, he traveled to Argentina, in South America, and while he was there he learned to sail and dance. He also visited poor villages where very few of the children had shoes. Mycoskie decided he had to do something.

"I'm going to start a shoe company, and for every pair I sell, I'm going to give one pair to a kid in need."

For this new **undertaking**, Mycoskie started the business using his own money. He named it TOMS: Shoes for Tomorrow. The slip-on shoes are modeled on shoes that are traditionally worn by Argentine workers.

Mycoskie immediately set up his **innovative** one-for-one program. TOMS gives away one pair of shoes for every pair that is purchased. Later that year, Mycoskie returned to Argentina and gave away 10,000 pairs of shoes. By 2011, TOMS had donated over one million pairs.

(bkgc) Kwaku Alston/Stockland Martel; (tr) Ho/Toms Shoe/AP Images

79

TOMS' employees unpack shoes to give away.

The company has expanded to sell eyeglasses. In a similar program, one pair of eyeglasses is donated for every pair that is bought.

Mycoskie is pleased and surprised. "I always thought I would spend the first half of my life making money and the second half giving it away," Mycoskie says. "I never thought I could do both at the same time."

Giving Back Rocks!

Have you ever seen a Hard Rock Cafe? The company runs restaurants and hotels. In 1990, the company launched a new **enterprise**: charity. Since then, it has given away millions of dollars to different causes. Its motto is Love All, Serve All.

One way the company raises **funds** for charity is by selling a line of T-shirts. The **process** starts with rock stars designing the art that goes on the shirts. Then the shirts are sold on the Internet. Part of the money that is raised from the sales of the shirts is given to charity.

Employees at Hard Rock Cafe locations are encouraged to raise money for their community. Every store does it differently.

The Hard Rock Cafes are successful and give back to the community.

Top Five Biggest Charities

| | $3,842,018,486 | $1,719,268,000 | $1,194,350,692 | $1,191,965,901 | $1,076,390,140 |

1 United Way Worldwide   **2 The Salvation Army**   **3 AmeriCares**   **4 Feed the Children**   **5 Food for the Poor**

Source: The Chronicle Of Philanthropy

Individuals as well as businesses are committed to helping people in need. This graph shows the American charities that got the most donations in one recent year and how much money they raised.

The restaurant in Hollywood, Florida, worked with some **exceptional** students from two Florida high schools. Together, they put on an event to raise money for the Make-A-Wish Foundation. The foundation grants wishes to children with serious medical problems.

The Bottom Line

Every day companies are thinking of innovative ways to give back to their community. If you own a business, making a profit is important. However, helping others is just as important as the bottom line. Helping others is good business!

Make Connections

How do the two companies profiled in this article help others? **ESSENTIAL QUESTION**

If you owned a business, how would you use some of your profits to help others? **TEXT TO SELF**

Reread

When you read an informational text, you may come across ideas and information that are new to you. As you read "Dollars and Sense," reread sections to make sure you understand the key facts and details in the text.

🔍 Find Text Evidence

As you read, you may want to make sure you understand the ways a business can help others. Reread the section "Hearts and Soles" in "Dollars and Sense."

page 79

G ood business is not always about the bottom line. A **compassionate** company knows that making money is not the only way to measure success. Many large businesses in the United States and all over the world are finding unusual ways to help people in need.

Hearts and Soles

After starting and running four businesses, Blake Mycoskie wanted a break from his usual **routine**. In 2006, he traveled to Argentina, in South America, and while he was there he learned to sail and dance. He also visited poor villages where very few of the children had shoes. Mycoskie decided he had to do something.

"I'm going to start a shoe company, and for every pair I sell, I'm going to give one pair to a kid in need."

For this new **undertaking**, Mycoskie started the business using his own money. He named it TOMS: Shoes for Tomorrow. The slip-on shoes are modeled on shoes that are traditionally worn by Argentine workers.

Mycoskie immediately set up his **innovative** one-for-one program. TOMS gives away one pair of shoes for every pair that is purchased. Later that year, Mycoskie returned to Argentina and gave away 10,000 pairs of shoes. By 2011, TOMS had donated over one million pairs.

> *I read that TOMS gives one pair of shoes for every pair of shoes someone buys. From this text evidence, I can draw the inference that the more shoes TOMS sells, the more shoes can be given away.*

Your Turn

COLLABORATE

What is another example of a company giving back to the community? Reread page 80 to answer the question. As you read other selections, remember to use the strategy Reread.

Main Idea and Key Details

The main idea is the most important idea that an author presents in a text or a section of text. Key details give important information to support the main idea.

 ### Find Text Evidence

When I reread the second paragraph in the section "Giving Back Rocks!" on page 80 of "Dollars and Sense," I can identify the key details. Next I can think about what the details have in common. Then I can figure out the main idea of the section.

Main Idea
Hard Rock Cafe sells a line of T-shirts to raise funds for charity.

Key details tell about the main idea.

Detail
Rock stars design the art that goes on the shirts.

Detail
The shirts are sold on the Internet.

Detail
Part of the money that is raised from the sales of the shirts is given to charity.

Your Turn

COLLABORATE

Reread the section "Hearts and Soles" on pages 79–80 of "Dollars and Sense." Find the key details in the section and list them in your graphic organizer. Use the details to determine the main idea.

Go Digital!
Use the interactive graphic organizer

Persuasive Article

"Dollars and Sense" is a persuasive article.

A persuasive article:

- Is nonfiction.
- States the writer's opinion on a topic.
- Provides facts and examples.
- May include text features such as headings and graphs.

Find Text Evidence

"Dollars and Sense" is a persuasive article. It states the author's opinion and tries to get readers to agree. It includes headings and a graph that shows the amount of money raised by different charities.

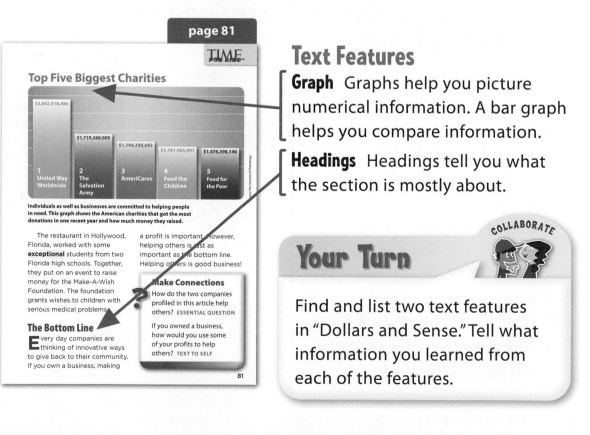

page 81

TIME FOR KIDS

Top Five Biggest Charities

$3,842,018,486 — 1 United Way Worldwide
$1,719,268,000 — 2 The Salvation Army
$1,194,350,692 — 3 AmeriCares
$1,191,965,901 — 4 Feed the Children
$1,076,390,140 — 5 Food for the Poor

Individuals as well as businesses are committed to helping people in need. This graph shows the American charities that got the most donations in one recent year and how much money they raised.

The restaurant in Hollywood, Florida, worked with some **exceptional** students from two Florida high schools. Together, they put on an event to raise money for the Make-A-Wish Foundation. The foundation grants wishes to children with serious medical problems.

a profit is important. However, helping others is just as important as the bottom line. Helping others is good business!

The Bottom Line

Every day companies are thinking of innovative ways to give back to their community. If you own a business, making

Make Connections

How do the two companies profiled in this article help others? ESSENTIAL QUESTION

If you owned a business, how would you use some of your profits to help others? TEXT TO SELF

81

Text Features

Graph Graphs help you picture numerical information. A bar graph helps you compare information.

Headings Headings tell you what the section is mostly about.

COLLABORATE

Your Turn

Find and list two text features in "Dollars and Sense." Tell what information you learned from each of the features.

Suffixes

A suffix is a word part added to the end of a word to change its meaning. Knowing some common suffixes can help you to figure out the meanings of unfamiliar words. Look at the suffixes below:

-ly = done in the way of

-ive = related or belonging to

-ful = full of or characterized by

Find Text Evidence

I see the word innovative *on page 79 of "Dollars and Sense." Looking at its word parts, I see the root word* innovate. *The suffix* -ive *changes a word into an adjective. This will help me to figure out what* innovative *means.*

Mycoskie immediately set up his innovative one-for-one program. TOMS gives away one pair of shoes for every pair that is purchased.

Your Turn

COLLABORATE

Use suffixes and context clues to figure out the meanings of the following words:

immediately, *page 79*

traditionally, *page 79*

successful, *page 80*

Readers to ...

Writers often vary the length of their sentences. A writer might follow a long sentence with a short sentence to draw attention to an important idea. Reread the paragraph from "Dollars and Sense" below.

Sentence Lengths

Identify long and short sentences. How do some of the shorter sentences draw attention to an idea?

Expert Model

After starting and running four businesses, Blake Mycoskie wanted a break from his usual routine. In 2006, he traveled to Argentina, in South America, and while he was there he learned to sail and dance. He also visited poor villages where very few of the children had shoes. Mycoskie decided he had to do something. "I'm going to start a shoe company, and for every pair I sell, I'm going to give one pair to a kid in need."

Writers

Courtney wrote about her favorite sport. Read Courtney's revisions to a section of her essay.

Grammar Handbook

Run-on Sentences
See page 454.

Student Model

Soccer Rules!

My favorite Sport to play is soccer. Soccer is a great sport to play because the action never stops! Last year, I joined a soccer team that travels all over the state to play in competitions. My position on the team is striker. it's my job to score as many goals as I can. When I score a goal, everyone around the field cheers. it's such a great feeling!

Your Turn

☑ Describe how Courtney's sentence lengths varied.
☑ Identify the run-on sentences and fragments she corrected.
☑ Tell how other revisions improved her writing.

Go Digital!
Write online in Writer's Workspace

Amazing Animals

A mouse in her room woke Miss Dowd;
She was frightened and screamed very loud,
Then a happy thought hit her—
To scare off the critter,
She sat up in bed and meowed.

— Anonymous

There was a young lady of Niger
Who smiled as she rode on a tiger;
They returned from the ride
With the lady inside,
And the smile on the face of the tiger.

— Cosmo Monkhouse

There was an Old Man with a beard,
Who said, "It is just as I feared!
Two Owls and a Hen,
Four Larks and a Wren,
Have all built their nests in my beard!"

— Edward Lear

The Big Idea

What can animals teach us?

? Essential Question
What are some messages in animal stories?

Go Digital!

ANIMAL TALES

Look at the photograph. If you wrote a story about this squirrel, what words would you use to describe him? Is he brave? Greedy? Clever? Foolish?

▶ What would the message of the story be?

▶ What are some animal stories you know that teach a lesson?

Talk About It

Write words you have learned about messages in animal stories. Talk with a partner about how animal stories show how people should behave.

Messages

Vocabulary

Use the picture and the sentences to talk with a partner about each word.

attracted

The brightly colored flower **attracted** a butterfly.

What kinds of insects are attracted to sugar?

dazzling

The fireworks in the night sky were bright and **dazzling**.

What is something else that can be described as dazzling?

fabric

The girl's towel and clothes are made from **fabric**.

What else can be made from fabric?

greed

It was **greed** that caused the boys to grab more cookies than they needed.

What is an antonym for greed?

honest

Leo was **honest** and told his mother the truth about the broken window.

What is a synonym for honest?

requested

The customer **requested** service from the waiter.

What is something you have requested?

soared

The seagull **soared** upward, high over the ocean.

What is a synonym for soared?

trudged

The tired hikers **trudged** slowly up the path.

Describe a time you trudged instead of walked quickly.

Your Turn

COLLABORATE

Pick three words. Write three questions for your partner to answer.

Go Digital! *Use the online visual glossary*

The Fisherman and the Kaha Bird

Essential Question

What are some messages in animal stories?

Read how a poor fisherman is helped by the Kaha bird.

Amanda Hall

ong ago there lived an old fisherman who made his pitiful living catching fish. All day the old man sat on the riverbank waiting for the fish to bite. But he never had more than one or two small fish to sell at the market. He and his wife were always hungry.

One morning, the tired old fisherman **trudged** slowly to the river. Suddenly a great bird with bright, **dazzling** silver feathers settled in the tree above him. The delighted fisherman knew this was the magnificent Kaha, a glittering bird that occasionally appeared to help the poor or the sick.

"I see you work for very little reward," the Kaha said. "I wish to help. Every day I will bring a large fish to your house that you can sell at the market."

The old man couldn't believe his luck. As the days passed, the **honest** Kaha kept her promise. The fisherman sold the fish and came home with plenty of food. Soon he was bringing home clothing made from brightly colored silk **fabric** for his wife.

At the market one day, the Shah's crier made an announcement: "Find the great Kaha for our eminent Shah and receive a reward of fifty bags of gold!"

The fisherman thought, "If I had fifty bags of gold, I would be rich! But how can I betray the bird?" He argued with himself until, finally, his **greed** for gold blinded him to the generosity of the Kaha bird.

He told the Shah's crier about the Kaha and **requested** assistance in catching her. He asked for four hundred men to help him.

That evening, four hundred servants followed the fisherman home. They hid among the trees as the fisherman set out a feast to tempt the bird. When the Kaha landed in a tree, the old man said, "Come dine with me, dear friend. I wish to express my gratitude."

The Kaha, touched by the fisherman's kindness and **attracted** to the delicious meal, flew down to join him. Immediately, the fisherman grabbed the Kaha by the feet and cried out to the servants to help him. The startled Kaha spread her wings. She began to fly up with the fisherman pulling at her. A servant caught the fisherman by the feet, but the bird rose higher. A second and third servant grabbed onto the first until soon four hundred servants hung by one another's feet as the Kaha **soared** upward.

Looking down, the fisherman could just barely see the river below. If he hadn't betrayed the Kaha, he would not be in this predicament. There was but one thing to do. The fisherman let go of the great bird's feet. The servants and the fisherman tumbled from the sky and landed in the river.

It was many weeks before the fisherman had healed enough to fish again. Every day the old fisherman looked up at the sky for a sign of the beautiful silver bird, but the Kaha was never seen again.

Make Connections

Talk about the message in this story. **ESSENTIAL QUESTION**

What would you tell the fisherman to convince him not to betray the Kaha bird? **TEXT TO SELF**

Amanda Hall

Ask and Answer Questions

When you read a story, you can ask questions before, during, and after you read to help you understand the story. As you read "The Fisherman and the Kaha Bird," look for answers to questions you may have about the story.

🔍 Find Text Evidence

You may want to know why the fisherman is poor. Reread the first paragraph of "The Fisherman and the Kaha Bird" on page 95 to find the answer.

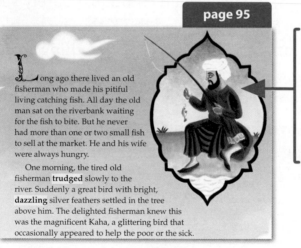

page 95

L͟ong ago there lived an old fisherman who made his pitiful living catching fish. All day the old man sat on the riverbank waiting for the fish to bite. But he never had more than one or two small fish to sell at the market. He and his wife were always hungry.

One morning, the tired old fisherman **trudged** slowly to the river. Suddenly a great bird with bright, **dazzling** silver feathers settled in the tree above him. The delighted fisherman knew this was the magnificent Kaha, a glittering bird that occasionally appeared to help the poor or the sick.

I read that the fisherman catches one or two small fish a day. Therefore, he does not make much money when he sells them.

 **Your Turn**

COLLABORATE

Reread the first two pages of "The Fisherman and the Kaha Bird" and list two questions you have about the Kaha bird. Find the answers in the text.

Theme

The theme is the central message or lesson that an author wants you to understand. To identify the theme of a story, look closely at what the characters say and do.

 Find Text Evidence

As I reread pages 95–96 of "The Fisherman and the Kaha Bird," I can look for clues to the theme by looking at the characters' words and the events of the story.

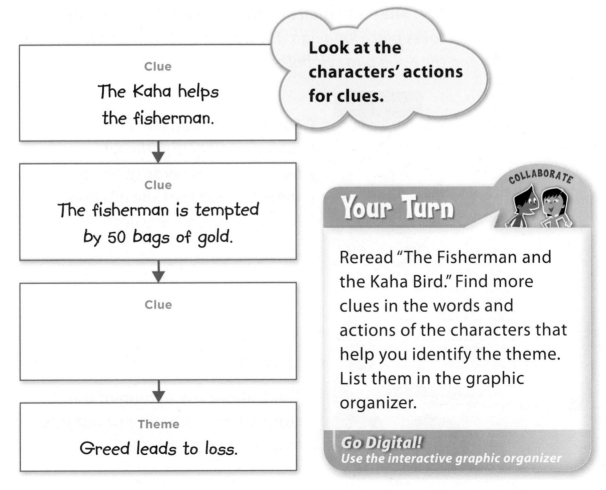

Clue

The Kaha helps the fisherman.

Look at the characters' actions for clues.

Clue

The fisherman is tempted by 50 bags of gold.

Clue

Theme

Greed leads to loss.

COLLABORATE

Your Turn

Reread "The Fisherman and the Kaha Bird." Find more clues in the words and actions of the characters that help you identify the theme. List them in the graphic organizer.

Go Digital!
Use the interactive graphic organizer

99

Folktale

The selection "The Fisherman and the Kaha Bird" is a folktale.

Folktales:

- Are based on the traditions and beliefs of a people.
- Usually teach a lesson.
- Often use animal characters to symbolize or represent a human quality.

Find Text Evidence

I can tell that "The Fisherman and the Kaha Bird" is a folktale. The story takes place long ago. The people believe in the Kaha bird, and the Kaha bird symbolizes a human quality.

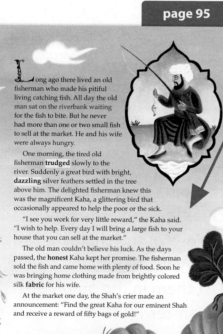

page 95

Symbolism *I can tell that the human quality that the Kaha represents is generosity. She tells the fisherman that she will bring him a large fish every day.*

Your Turn

Think about the ending of the story. Discuss whether or not this folktale teaches a lesson.

Root Words

As you read "The Fisherman and the Kaha Bird," you may come across a word that you don't know. Look for the simplest form of the word, the **root**. This will help you figure out the meaning of the longer word.

 Find Text Evidence

On page 95 in "The Fisherman and the Kaha Bird," I am not sure what the word announcement *means. I see that the word contains the root word* announce. *This will help me to figure out what the word* announcement *means.*

> At the market one day, the Shah's crier made an announcement: "Find the great Kaha for our eminent Shah and receive a reward of fifty bags of gold!"

Your Turn

COLLABORATE

With a partner, use root words to find the meanings of these words from "The Fisherman and the Kaha Bird":

    pitiful, *page 95*

    generosity, *page 96*

    assistance, *page 96*

Write a short definition and an example sentence for each word.

Amanda Hall

Readers to...

Writers include a strong opening when writing a narrative. A strong beginning hooks the reader's attention and gives clues about what will happen next. Reread the excerpt from "The Fisherman and the Kaha Bird" below.

Strong Openings

The author creates a **strong opening** to the story. What details in the first two paragraphs make you want to read more?

Expert Model

The Fisherman and the Kaha Bird

Long ago there lived an old fisherman who made his pitiful living catching fish. All day the old man sat on the riverbank waiting for the fish to bite. But he never had more than one or two small fish to sell at the market. He and his wife were always hungry.

One morning, the tired old fisherman trudged slowly to the river. Suddenly a great bird with bright, dazzling silver feathers settled in the tree above him.

Amanda Hall

Writers

Editing Marks

⌐⌐ Switch order.

∧ Add.

⌄ Add a comma.

✓ Take out.

(SP) Check spelling.

≡ Make a capital letter.

Lina wrote a story. Read Lina's revision of the beginning of her story.

Grammar Handbook

Common and Proper Nouns See page 456.

Student Model

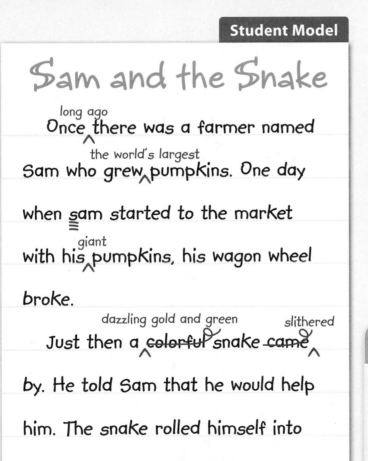

Sam and the Snake

                    long ago
Once ∧ there was a farmer named

        the world's largest
Sam who grew ∧ pumpkins. One day

when ṣam started to the market

            giant
with his ∧ pumpkins, his wagon wheel

broke.

        dazzling gold and green   slithered
Just then a ~~colorful~~ snake ~~came~~ ∧

by. He told Sam that he would help

him. The snake rolled himself into

a wheel and attached himself to

the wagon.

Your Turn

COLLABORATE

☑ Identify the details of Lina's strong opening.
☑ Identify a proper noun she included.
☑ Tell how Lina's revisions improved her writing.

Go Digital!
Write online in Writer's Workspace

Essential Question
How do animal characters change familiar stories?

Go Digital!

Look Familiar?

Hi, I'm a handsome prince. I just happen to be a frog at the moment. My story is about courage and amazing transformations. Of course it ends happily.

▶ There are lots of great animal stories. What are some of your favorites?

▶ Not all animals are as smart or charming as me. What are some character traits of the animals in your favorite stories?

Talk About It

Write some of the words that describe traits of animals in stories. Talk about these traits with your partner.

Traits

Vocabulary

Use the picture and the sentences to talk with a
partner about each word.

annoyed

Having to wake up early **annoyed** my
father and made him grumpy.

What is a synonym for annoyed?

attitude

The girls had fun working together
because they both had a good **attitude**.

Describe your attitude about doing
chores.

commotion

The swans made a **commotion** with
their squawking and splashing.

What is an example of something that
can make a commotion?

cranky

Being hungry makes Neil feel **cranky**.

What makes you feel cranky?

familiar

I took a **familiar** route from the bus stop to my house so that I would not get lost.

What is a familiar sound when you go to the park?

frustrated

The student was **frustrated** by the difficult assignment.

What type of situation makes you feel frustrated?

selfish

The two friends were not **selfish** at all and shared everything.

How would you describe a selfish person?

specialty

Understanding x-rays is a **specialty** that requires training and practice.

What kind of specialty would require training in how to fly a plane?

Your Turn

COLLABORATE

Pick three words. Write three questions for your partner to answer.

Go Digital! *Use the online visual glossary*

The Ant and the Grasshopper

? Essential Question

How do animal characters change familiar stories?

Read about how an ant teaches a grasshopper an important lesson.

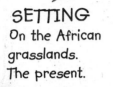

SETTING
On the African grasslands.
The present.

CHARACTERS
Termite (the narrator)

Ant

Grasshopper

An army of ants

Scene I

(It is raining heavily on the African grasslands. Termite turns and sees the audience.)

TERMITE: *(Happily)* Yipes! I didn't see you. Welcome to the great plains of Africa! We're soggy now because it's the rainy season. Sorry. *(She shrugs and smiles.)* Today, we'll visit two very different friends of mine— Ant and Grasshopper. Maybe you have heard of them from other **familiar** stories. Let's see what my buddies are up to!

(An army of ants march in, carrying leaves filled with water. They approach Grasshopper, who lounges lazily under a plant.)

ANT: *(In a loud voice)* Company, halt! *(The ants stop.)*

GRASSHOPPER: *(Stretching and yawning)* Ant, old pal! Good to see you! I was just napping when I heard your feet pounding down the way. What's all the **commotion**?

Emily Carew Woodard

109

ANT: (*Looking **annoyed***) Grasshopper, have you noticed what falls from the sky above you?

(*Ant stands at attention and points up at a cloud. Grasshopper sleepily rises and stands next to Ant. He looks at the sky.*)

ANT: Rain, Grasshopper! Rain falls from the sky! And when there is rain, there is work to be done.

GRASSHOPPER: (*Smiling then scratching his head*) Huh?

ANT: (*Sighing*) You should be collecting water for a time when it is unavailable. Instead, you lie here without a care for the future.

GRASSHOPPER: (*Laughing*) Oh, don't be so serious, ol' buddy! There is plenty of water now, and that's all that matters. You need to relax! You're much too tense. Why don't you make napping your new **specialty** instead of all this silly toil? Stop working so hard all the time!

ANT: (*Shaking his head as he grows **frustrated***) The rainy season will not last forever, Grasshopper. Your carefree **attitude** will disappear with the water, and soon you will regret being lazy and wish you had been more energetic.

(*The ants march off as Grasshopper continues to laugh.*)

Scene II

(*It is a few months later, and the plains are now dusty, dry, and brown. Grasshopper, appearing weak and sickly, knocks on Ant's door. Ant, seeming strong and healthy, opens the door.*)

Emily Carew Woodard

110

GRASSHOPPER: (*Nervously*) Hi there, pal…. I was in the neighborhood. Boy, can you believe how hot it is? So… uh…I was wondering if maybe…by chance…you might have some water for your old friend.

(*Ant tries to close the door, but Grasshopper quickly grabs it.*)

GRASSHOPPER: (*Begging wildly*) PLEASE, Ant! I am so thirsty! There isn't a drop of water anywhere!

ANT: (*After a pause*) We ants worked hard to collect this water, but we cannot let you suffer. (*Giving Grasshopper a sip of water*) Do not think us **selfish**, but we can only share a few drops with you. I warned you that this time would come. If you had prepared, you would not be in this situation.

(*Grasshopper walks slowly away. Termite watches him go.*)

TERMITE: Although Ant has done a good deed, tired, **cranky** Grasshopper must still search for water. Grasshopper learned an important lesson today. Next time, he will follow Ant's advice!

Make Connections

Talk about how Ant and Grasshopper act like real people. **ESSENTIAL QUESTION**

Explain why you are more like Ant or more like Grasshopper. **TEXT TO SELF**

Ask and Answer Questions

When you read a selection, you may not understand all of it. It helps to stop and ask yourself questions. As you read "The Ant and the Grasshopper," ask questions about what you don't understand. Then read to find the answers.

Find Text Evidence

After reading Scene I, you may ask yourself what happens in Africa after the rainy season ends. Reread Scene II of "The Ant and the Grasshopper" to find the answer.

page 110

season will not last forever, Grasshopper. Your carefree **attitude** will disappear with the water, and soon you will regret being lazy and wish you had been more energetic.

(The ants march off as Grasshopper continues to laugh.)

Scene II

(It is a few months later, and the plains are now dusty, dry, and brown. Grasshopper, appearing weak and sickly, knocks on Ant's door. Ant, seeming strong and healthy, opens the door.)

110

I read that the land is "dusty, dry, and brown." Grasshopper is weak. This suggests that there are long periods of time when no rain falls in Africa.

COLLABORATE

Your Turn

In "The Ant and the Grasshopper," what is Termite's role? Reread to answer this question. List two other questions you have about the story and read to find the answers.

Theme

The theme of a selection is the message or lesson that an author wants to communicate to the reader. To identify the theme, pay attention to the characters' words and actions.

 Find Text Evidence

As I reread "The Ant and the Grasshopper," the different actions of Ant and Grasshopper in the rainy season seem like important clues to the theme. So do Ant's words about collecting water.

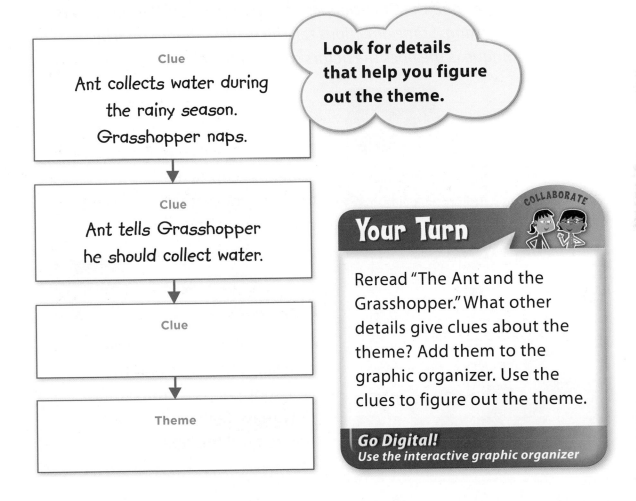

> **Clue**
>
> Ant collects water during the rainy season. Grasshopper naps.

Look for details that help you figure out the theme.

> **Clue**
>
> Ant tells Grasshopper he should collect water.

> **Clue**

> **Theme**

COLLABORATE

Your Turn

Reread "The Ant and the Grasshopper." What other details give clues about the theme? Add them to the graphic organizer. Use the clues to figure out the theme.

Go Digital!
Use the interactive graphic organizer

113

Drama

The fable "The Ant and the Grasshopper" is presented as a drama.

> **A drama:**
> - Has a list of characters and is written in dialogue.
> - Is divided into parts called *acts* or *scenes*.
> - Includes the setting and stage directions.

Find Text Evidence

"The Ant and the Grasshopper" is a drama. It lists the characters and setting. Stage directions tell the actors what to do. The dialogue is the lines the actors speak in the play.

page 109

Scene I

(It is raining heavily on the African grasslands. Termite turns and sees the audience.)

TERMITE: *(Happily)* Yipes! I didn't see you. Welcome to the great plains of Africa! We're soggy now because it's the rainy season. Sorry. *(She shrugs and smiles.)* Today, we'll visit two very different friends of mine— Ant and Grasshopper. Maybe you have heard of them from other **familiar** stories. Let's see what my buddies are up to!

(An army of ants march in, carrying leaves filled with water. They approach Grasshopper, who lounges lazily under a plant.)

ANT: *(In a loud voice)* Company, halt! *(The ants stop.)*

GRASSHOPPER: *(Stretching and yawning)* Ant, old pal! Good to see you! I was just napping when I heard your feet pounding down the way. What's all the **commotion**?

SETTING
On the African grasslands.
The present.

CHARACTERS
Termite (the narrator)
Ant
Grasshopper
An army of ants

109

> **Dialogue** The characters' names appear in capital letters before the lines they speak.

Your Turn

COLLABORATE

Find and list two other examples that show "The Ant and the Grasshopper" is a drama. Tell how these features help you understand the text.

Antonyms

As you read "The Ant and the Grasshopper," you may come across a word you don't know. Sometimes the author will use an **antonym**, another word or phrase that means the opposite of the unfamiliar word.

 Find Text Evidence

On page 110 of "The Ant and the Grasshopper," I'm not sure what carefree *means. I can use the word* serious *to help me figure out what* carefree *means.*

"Oh, don't be so serious, ol' buddy! There is plenty of water now, and that's all that matters. You need to relax!"

COLLABORATE

Your Turn

Use antonyms to figure out the meanings of the words below in "The Ant and the Grasshopper."

tense, *page 110*
energetic, *page 110*
sickly, *page 110*

Emily Carew Woodard

Readers to...

Writers often use an informal voice when writing a play. They use everyday words and phrases that sound like conversation. Reread Termite's greeting from "The Ant and the Grasshopper" below.

Informal Voice

Identify words and phrases that show an **informal voice**. What can you tell about Termite from her words?

Expert Model

The Ant and the Grasshopper

TERMITE: (*Happily*) Yipes! I didn't see you. Welcome to the great plains of Africa! We're soggy now because it's the rainy season. Sorry. (*She shrugs and smiles.*) Today, we'll visit two very different friends of mine—Ant and Grasshopper. Maybe you have heard of them from other familiar stories. Let's see what my buddies are up to!

Writers

Editing Marks

⌐⌐ Switch order.

∧ Add.

∧ Add a comma.

⌀ Take out.

(SP) Check spelling.

≡ Make a capital letter.

Sophie wrote a story. Read Sophie's revision of the beginning of her story.

Grammar Handbook

Singular and Plural Nouns See page 455.

Student Model

The Hare's Side

guess I
I∧should introduce myself first.

I am the Hare. The one that Tortoise

supposedly
beat in a race. The story is that
∧

bragged
I said I was fast and made fun of
∧

Tortoise. Not true! Ask all my friend;
∧

they'll tell the real story. The whole

thing was a setup⊙ I was tired that
∧

day from running errands. Any other

day, I'd have won that race no sweat.

Your Turn

COLLABORATE

☑ Identify examples of informal voice in Sophie's story.

☑ Identify two plural nouns in her story.

☑ Tell how revisions improved her writing.

Go Digital!
Write online in Writer's Workspace

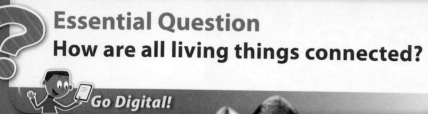

Essential Question
How are all living things connected?

Go Digital!

PART OF THE WHOLE

The relationship between the oxpecker and the zebra is one example of the relationships that exist in the animal world. These birds eat ticks and lice off zebras. They also screech loudly if a predator is approaching.

▶ What are some other examples of relationships between animals?

▶ How are animals and plants connected?

Talk About It

COLLABORATE

Write words that describe how living things are connected. Then talk to a partner about why these connections are so important.

Connections

(tkgd) Elvele Images Ltd./Alamy; (tt, bc) John Foxx/Stockbyte Silver/Getty Images; (tt, br, tl, bl) Siede Preis/Photodisc/Getty Images

Vocabulary

Use the picture and the sentences to talk with a partner about each word.

crumbled

The old brick wall had **crumbled** over the years.

What is a synonym for crumbled?

droughts

Because of the lack of rain, farmers' crops died during the **droughts**.

In what part of the world are there a lot of droughts?

ecosystem

A reef **ecosystem** can be disrupted if you remove one species that lives in it.

What are some other examples of ecosystems?

extinct

The American buffalo was hunted so much that it almost became **extinct**.

Name an animal that is now extinct.

flourished

The sunflowers grew tall and **flourished** in the rich soil.

What is a synonym for flourished?

fragile

Tom held the nest carefully because he was afraid the **fragile** eggs might break.

What is an antonym for fragile?

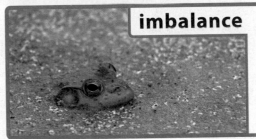

imbalance

Too much algae created an **imbalance** in the pond's ecosystem.

How are the words imbalance and inequality similar?

ripples

The water **ripples** around the swimming dog.

If a flag ripples, is the air windy or still?

COLLABORATE

Your Turn

Pick three words. Write three questions for your partner to answer.

Go Digital! *Use the online visual glossary*

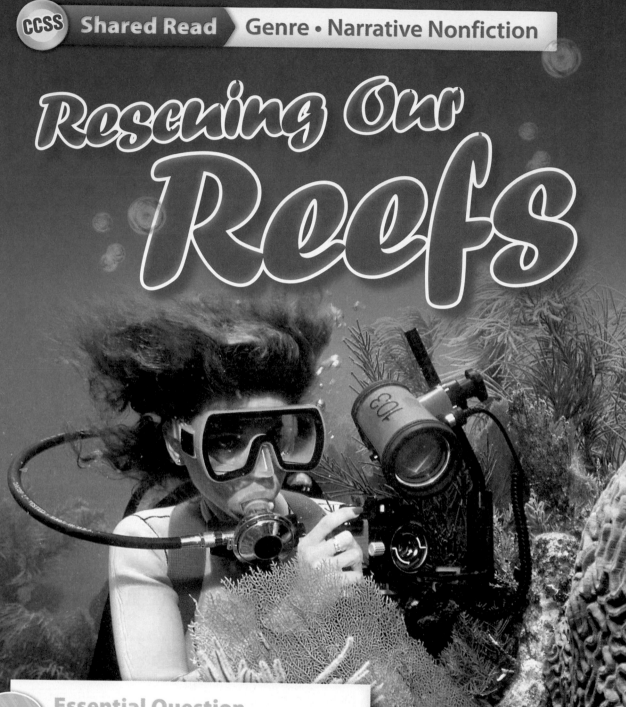

Rescuing Our Reefs

Essential Question

How are all living things connected?

Read how plants and animals are connected in a coral reef ecosystem.

Sitting on the side of the boat, the photographer fixes her scuba tank and mask. She waves to a man in a fishing boat. Then she dives backwards into the clear waters of the Florida Keys. She swims, breathing through her regulator. A large, colorful coral reef is laid out before her eyes. Sea anemones, red hind fish, gaudy parrotfish, yellow angelfish, and other animals ignore her as they go about their business. Life in this reef has flourished and grown.

Connections

The photographer knows the plants and animals in a reef ecosystem need each other to survive. Reefs are made up of billions of tiny animals called coral polyps. Plant-like algae live inside the coral. The algae use a process called photosynthesis to turn energy from the sun into food for themselves and the coral. In return, the coral gives the algae a home and the carbon dioxide needed for photosynthesis. Algae are a part of the food chain called producers. Producers make their own energy.

The photographer sees a blue and yellow parrotfish nibbling at the coral. She takes a picture. The parrotfish breaks apart the coral to get to the algae-filled polyps inside. In a food chain, the parrotfish is a consumer. Consumers cannot produce their own energy. As the parrotfish eats the algae, energy is passed through the food chain.

Parrotfish

In the distance, the photographer notices the long silver body of a barracuda lurking. The sea grass ripples in the current, swaying back and forth. It almost hides the hungry predator. She snaps a photo and swims on.

(bkgd) Stephen Frink/Corbis; (inset) Timothy Grollimund

Coral Bleaching

The photographer shoots more photos as she swims. The reef must have looked like this hundreds of years ago. But then she stops and stares at a big area of bleached, white coral. Once colorful, the whitish coral now looks like the broken pieces of a **crumbled** castle.

Coral depends on a natural balance to stay healthy. Climate change and pollution can cause an **imbalance**. Some areas have dried up from **droughts** while others have had more rain. Too much sun and warmer ocean temperatures can cause coral bleaching.

If pollution gets into the water or the water gets too warm, the relationship between the coral and algae breaks down. The algae stop making food. The coral ejects the algae. The algae are what give the coral its color. The coral loses its color. It starves because it needs the algae to make food for it.

A food chain shows the transfer of energy from one species to another.

Energy Source **Producer** **Consumer**

Many plants and animals depend on the coral for food and shelter. As more and more coral reefs die, many animals and plants that live in these reefs may become extinct. The beautiful reef the photographer had seen earlier would resemble the white, crumbling reef before her.

Balancing Act

She turned and swam back to the boat. Later today, she would send her photographs to the Nature Conservancy. It is an organization that works to rescue our fragile reefs. Scientists there are trying to rebuild the reefs by attaching small pieces of staghorn coral to concrete blocks. Staghorn coral is used to grow new coral. Once the coral grows, the blocks are planted in the reefs.

The photographer hopes her pictures will help spread the word. They show the relationship between pollution, climate change, and coral bleaching. She breaks through the water's surface and climbs into the boat.

"I got some good shots of the healthy reef and the sick reef!" she shouts to her partner. Once aboard, she immediately begins putting her photos on her laptop.

Consumer

Make Connections

Talk about how the plants and animals that live in the coral reef are connected. **ESSENTIAL QUESTION**

What could you do to help save the coral reefs? **TEXT TO SELF**

(b) Richard Carey/Alamy

Summarize

When you summarize, you retell the most important details in a paragraph or section of text. Summarize sections of "Rescuing Our Reefs" to help you understand the information.

Find Text Evidence

As you read, identify the most important details. Summarize the first paragraph of the "Connections" section on page 123 of "Rescuing Our Reefs."

page 123

Sitting on the side of the boat, the photographer fixes her scuba tank and mask. She waves to a man in a fishing boat. Then she dives backwards into the clear waters of the Florida Keys. She swims, breathing through her regulator. A large, colorful coral reef is laid out before her eyes. Sea anemones, red hind fish, gaudy parrotfish, yellow angelfish, and other animals ignore her as they go about their business. Life in this reef has flourished and grown.

Connections

The photographer knows the plants and animals in a reef ecosystem need each other to survive. Reefs are made up of billions of tiny animals called coral polyps. Plant-like algae live inside the coral. The algae use a process called photosynthesis to turn energy from the sun into food for themselves and the coral. In return, the coral gives the algae a home and the carbon dioxide needed for photosynthesis. Algae are a part of the food chain called producers. Producers make their own energy.

In a coral reef ecosystem, the algae and coral polyps help each other. The algae produces food through photosynthesis, and the coral provide carbon dioxide and a home for the algae.

Your Turn

COLLABORATE

Reread "Coral Bleaching" on pages 124–125. Summarize the third paragraph. As you read, remember to use the strategy Summarize.

Main Idea and Details

The main idea is the most important idea that an author presents in a paragraph or section of text. Key details give important information to support the main idea.

 Find Text Evidence

When I reread the section "Connections" on page 123 in "Rescuing Our Reefs," I can reread to find the key details. Then I can think about what those details have in common. Now I can figure out the main idea of the section.

All the key details tell about the main idea.

Main Idea
Animals and plants in the coral reef depend on each other.

Detail
Algae produce food through the process of photosynthesis.

Detail
The coral provides a home and carbon dioxide for the algae.

Detail
Parrotfish eat the algae inside the coral.

Your Turn
COLLABORATE

Reread "Coral Bleaching" on pages 124–125. Find the key details and list them in your graphic organizer. Use the details to find the main idea.

Go Digital!
Use the interactive graphic organizer

Narrative Nonfiction

"Rescuing Our Reefs" is narrative nonfiction.

Narrative nonfiction:

- Tells a story.
- Presents facts about a topic.
- Includes text features.

Find Text Evidence

"Rescuing Our Reefs" is narrative nonfiction. It tells a story while providing facts about reefs. It also includes text features.

---

page 124

Coral Bleaching

The photographer shoots more photos as she swims. The reef must have looked like this hundreds of years ago. But then she stops and stares at a big area of bleached, white coral. Once colorful, the whitish coral now looks like the broken pieces of a crumbled castle.

Coral depends on a natural balance to stay healthy. Climate change and pollution can cause an imbalance. Some areas have dried up from droughts while others have had more rain. Too much sun and warmer ocean temperatures can cause coral bleaching.

If pollution gets into the water or the water gets too warm, the relationship between the coral and algae breaks down. The algae stop making food. The coral ejects the algae. The algae are what give the coral its color. The coral loses its color. It starves because it needs the algae to make food for it.

A food chain shows the transfer of energy from one species to another.

Energy Source — Producer — Consumer

124

Text Features

Headings Headings tell what a section of text is mostly about.

Flow Chart A flow chart shows information from the text in a visual way.

Your Turn

Find and list two examples of text features in "Rescuing Our Reefs." Tell your partner what information you learned from each of the features.

Context Clues

As you read "Rescuing Our Reefs," you may come across words you don't know. To figure out the meaning of an unfamiliar word, use the words, phrases, and sentences near it for clues.

Find Text Evidence

On page 123 of "Rescuing Our Reefs," I see that the narrator says, "In a food chain, the parrotfish is a consumer." I'm not sure what a consumer *is. I read the next sentence, "Consumers cannot produce their own energy." Now I know what* consumer *means.*

In a food chain, the parrotfish is a consumer. Consumers cannot produce their own energy.

Your Turn

Use context clues to find the meanings of the words below in "Rescuing Our Reefs." Write a short definition and an example sentence for each word.

predator, *page 123*
bleached, *page 124*
ejects, *page 124*

Timothy Grollimund

Readers to...

Writers include specific details to support their main ideas. They provide facts that tell more about the information they want to share. Reread the excerpt from "Rescuing Our Reefs" below.

Expert Model

Supporting Details

Identify the **supporting details.** How do these details support the main idea?

The photographer sees a blue and yellow parrotfish nibbling at the coral. She takes a picture. The parrotfish breaks apart the coral to get to the algae-filled polyps inside. In a food chain, the parrotfish is a consumer. Consumers cannot produce their own energy. As the parrotfish eats the algae, energy is passed through the food chain.

Writers

Carlos wrote an article. Read Carlos's revisions of a section of his article.

SEA ANEMONES

Sea anemones and clownfishes~~es~~ live in coral reefs. They need each other to survive. The clownfish protects the sea anemone. It also ^by scaring away predators^ cleans the anemone and feeds it. ^by dropping scraps of food for the anemone to eat^ In return, the sea anemone gives the clownfish a home in its ^poisonous^ tentacles. The clownfish is immune to the anemone's poison. Both the sea anemone and the clownfish need the coral reef, too.

The coral reef is their ecosystem.

Your Turn

- ✔ Identify the supporting details in Carlos's article.
- ✔ Identify an irregular plural noun that he used.
- ✔ Tell how Carlos's revisions help support the main idea.

Go Digital!
Write online in Writer's Workspace

Essential Question
What helps an animal survive?

Go Digital!

Adapting to Survive

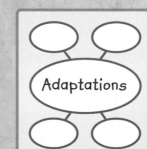

Hi, I'm a chameleon. Have you ever seen anyone quite like me? Here's how I have adapted to survive in my environment.

▶ See my skin color? I can change it. Changing my skin color helps to camouflage me from predators.

▶ My long tail can wrap around branches. How do you think that helps me?

Talk About It COLLABORATE

Write words you have learned about adaptation. Then talk with a partner about other animals and how they have adapted to survive.

Adaptations

Vocabulary

Use the picture and the sentences to talk with a partner about each word.

camouflaged

It is hard to see the **camouflaged** insect because it blends in with the leaf.

How are the words camouflaged and hidden similar?

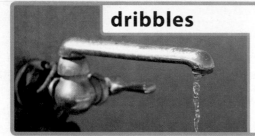

dribbles

Water **dribbles** from the leaky faucet all night.

Name something else that dribbles.

extraordinary

The owl has an **extraordinary** ability to stare for a long time without blinking.

What is an antonym for extraordinary?

poisonous

Some wild mushrooms can make you sick because they are **poisonous**.

What other things are poisonous?

pounce

The bobcat likes to **pounce** on fish in the river.

What other animals pounce?

predator

A leopard is a fierce **predator** that can catch most animals that it hunts.

Explain why a hawk is a predator.

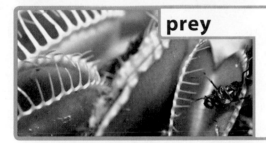

prey

The venus flytrap plant traps its **prey** inside its sticky leaves.

What is an antonym for prey?

vibrations

Eric plucked his guitar strings, causing **vibrations** as the strings moved quickly back and forth.

What else can make vibrations?

COLLABORATE

Your Turn

Pick three words. Write three questions for your partner to answer.

Go Digital! *Use the online visual glossary*

Animal
Adaptations

Essential Question

What helps an animal survive?

Read about ways different animals adapt to their environments.

What would you do if you saw a skunk raise his tail? If you knew anything about skunks, you would run in the opposite direction! Skunks have a built-in survival system. They can blast a **predator** with a horrible-smelling spray produced by the glands under their tails.

The special ways that animals have to survive are called adaptations. These include physical traits such as the skunk's spray and animals with bright colors and markings that warn predators that they are **poisonous**. Some animals can sense the smallest **vibrations** in the ground. Others hear sounds from miles away. An adaptation can also be a behavioral trait. An example of a behavioral trait would be birds that migrate south every winter to avoid harsh temperatures.

When a skunk turns and sprays a predator, the foul-smelling mist can travel up to 10 feet.

Staying Warm

Brrrr! Imagine living in a place where the average annual temperature is an **extraordinary** 10 to 20° F. Welcome to the Arctic tundra of Alaska, Canada, Greenland, and Russia, home of the caribou. To stay warm, caribou have two layers of fur and a thick layer of fat. They also have compact bodies. Only 4 or 5 feet long, caribou can weigh over 500 pounds.

The tip of the caribou's nose and mouth is called a muzzle. It is covered in short hair. This hair helps to warm the air before they inhale it into their lungs. It also helps to keep them warm as they push snow aside to find food.

Finding Food

Every day, a caribou eats over six pounds of lichen! Caribou have unusual stomachs. The stomach's four chambers are designed to digest lichen. It is one of the few foods they can find in the winter. Even so, caribou still have a tough time in the coldest part of winter when their food sources decline. That's why they travel from the tundra to a large forest area, where food is easier to find. When the melting snow **dribbles** into streams, they know that it is time to return up north.

Lichen can grow in extreme temperatures.

Insects in Disguise

Look closely at the photo of the tree branch. Can you spot the insect? It is a phasmid. Some phasmids are known as leaf insects, or walking sticks. Phasmids look like leaves or twigs. These insects can change colors to really blend in with their surroundings. In this way, they are **camouflaged** from predators. It's as if they disappear from sight! These insects are nocturnal, which means that they are active at night. This is another adaptation that helps them avoid predators. It's hard to spot these insects in daylight, let alone at night.

This phasmid is called a walking stick because it looks like a stick with legs.

Water, Please!

In Florida's vast Everglades ecosystem, the dry season is brutal for many plants and animals. Alligators have found a way to survive these dry conditions in the freshwater marshes. They use their feet and snouts to clear dirt from holes in the limestone bedrock. When the ground dries up, the alligators can drink from their water holes.

Other species benefit from these water holes, too. Plants grow there. Other animals find water to survive the dry season. However, the animals that visit alligator holes become easy **prey**. The normally motionless alligator may **pounce** on them without warning. But luckily, alligators eat only a few times each month. Many animals take their chances and revisit the alligator hole when they need water. In the end, it's all about survival!

Make Connections

? How do adaptations help an animal survive? **ESSENTIAL QUESTION**

Describe an animal adaptation that you have seen. **TEXT TO SELF**

(bkgd) Pete Oxford/Minden Pictures

Summarize

When you summarize, you retell the most important details in a paragraph or section of text. Summarize sections of "Animal Adaptations" to help you understand the information.

Find Text Evidence

Reread the section "Insects in Disguise" on page 138. Identify key details to summarize the section.

page 138

Insects in Disguise

Look closely at the photo of the tree branch. Can you spot the insect? It is a phasmid. Some phasmids are known as leaf insects, or walking sticks. Phasmids look like leaves or twigs. These insects can change colors to really blend in with their surroundings. In this way, they are **camouflaged** from predators. It's as if they disappear from sight! These insects are nocturnal, which means that they are active at night. This is another adaptation that helps them avoid predators. It's hard to spot these insects in daylight, let alone at night.

This phasmid is called a walking stick because it looks like a stick with legs.

138

Phasmids are insects that can camouflage themselves to avoid predators. In addition, phasmids are nocturnal, which makes them difficult for predators to spot.

Your Turn

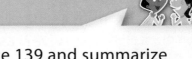
COLLABORATE

Reread "Water, Please!" on page 139 and summarize the section.

Main Idea and Key Details

The main idea is the most important point that the author makes in a text or a section of the text. Key details give important information to support the main idea.

 Find Text Evidence

When I reread the section "Staying Warm" in "Animal Adaptations" on page 137, I can identify the key details. Then I can think about what those details have in common. Now I can figure out the main idea of the section.

All three key details support the main idea.

Main Idea
Caribou adaptations help them survive the cold.

Detail
Caribou have two layers of fur and a thick layer of fat.

Detail
Short hair on their muzzles warms the air that they inhale.

Detail
Caribou have compact bodies that can weigh over 500 lbs.

Your Turn COLLABORATE

Reread the section "Insects in Disguise" on page 138. Find the key details and list them in the graphic organizer. Use these details to figure out the main idea.

Go Digital!
Use the interactive graphic organizer

Expository Text

"Animal Adaptations" is an expository text.

Expository text:
- Gives facts and information about a topic.
- Includes text features.

Find Text Evidence

"Animal Adaptations" is an expository text. It gives me facts about how different animals have adapted to survive. Each section has a heading. The text also includes photographs and captions.

page 138

Finding Food

Every day, a caribou eats over six pounds of lichen! Caribou have unusual stomachs. The stomach's four chambers are designed to digest lichen. It is one of the few foods they can find in the winter. Even so, caribou still have a tough time in the coldest part of winter when their food sources decline. That's why they travel from the tundra to a large forest area, where food is easier to find. When the melting snow **dribbles** into streams, they know that it is time to return up north.

Lichen can grow in extreme temperatures.

Insects in Disguise

Look closely at the photo of the tree branch. Can you spot the insect? It is a phasmid. Some phasmids are known as leaf insects, or walking sticks. Phasmids look like leaves or twigs. These insects can change colors to really blend in with their surroundings. In this way, they are **camouflaged** from predators. It's as if they disappear from sight! These insects are nocturnal, which means that they are active at night. This is another adaptation that helps them avoid predators. It's hard to spot these insects in daylight, let alone at night.

This phasmid is called a walking stick because it looks like a stick with legs.

138

Text Features

Photographs and Captions
Photographs illustrate what is in the text. Captions provide additional information.

Headings Headings tell what a section of text is mostly about.

Your Turn

COLLABORATE

Find and list two text features in "Animal Adaptations." Tell your partner what information you learned from each of these features.

Prefixes

As you read "Animal Adaptations," you may come across a word that you don't know. Look for word parts such as prefixes. A prefix is added to the beginning of a word and changes the meaning of the word. Here are some common prefixes.

un- means "not "

re- means "again"

dis- means "opposite of "

Find Text Evidence

When I read the section "Staying Warm" on page 137 in "Animal Adaptations," I see the word extraordinary. *First, I look at the separate word parts. I know that* extra *is a prefix that changes the meaning of* ordinary. *The prefix* extra *means "beyond. "*

Imagine living in a place where the average annual temperature is an extraordinary 10 to 20° F.

Your Turn

COLLABORATE

Use prefixes and context clues to figure out the meanings of the following words in "Animal Adaptations":

unusual, *page 138*

disappear, *page 138*

revisit, *page 139*

Readers to ...

Writers organize the information in expository text in a logical way. A compare-and-contrast text structure is one way to present information. Reread the excerpt from "Animal Adaptations" below.

Expert Model

Logical Order

Identify the **logical order** the details are presented in. Do they support the main idea?

The special ways that animals have to survive are called adaptations. These include physical traits such as the skunk's spray and animals with bright colors and markings that warn predators that they are poisonous. Some animals can sense the smallest vibrations in the ground. Others hear sounds from miles away. An adaptation can also be a behavioral trait. An example of a behavioral trait would be birds that migrate south every winter to avoid harsh temperatures.

Writers

Sonal wrote an expository text. Read Sonal's revisions of one section.

Student Model

PORCUPINES

A porcupine has special adaptations to help it survive in nature.

Porcupines move slow. They can't outrun their predators. This is unlike Rabbits can quickly hop away. A porcupines predators include coyotes and owls.

Unlike other animals, a porcupine has a powerful defense: its quills. When an animal attacks a porcupine, it releases its quills, and this is painful to a predator. Each quill has a barb.

Editing Marks

⌐⌐ Switch order.

∧ Add.

∧ Add a comma.

✌ Take out.

(sp) Check spelling.

≡ Make a capital letter.

Grammar Handbook

Possessive Nouns
See page 457.

Your Turn

COLLABORATE

☑ Identify the logical order that Sonal used.

☑ Identify a possessive noun that she used.

☑ Tell how Sonal's revisions improved her writing.

Go Digital!
Write online in Writer's Workspace

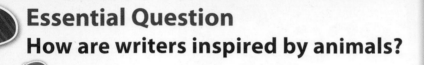

Essential Question
How are writers inspired by animals?

Go Digital!

Creative Leaps

Writers are observers. They watch and listen and then are inspired to create a picture of their experiences with words. Through the centuries, animals have provided writers with inspiration for countless stories, plays, and poems.

▶ What is your favorite story, play, or poem about an animal?

▶ What animal would you like to write about? Why?

Talk About It

Identify animal traits that might inspire a writer. Then talk with a partner about a favorite animal and explain what kind of story, play, or poem you might write about it.

Animal Traits

Vocabulary

Use the picture and the sentences to talk with a partner about each word.

brittle

The **brittle**, dry leaf fell apart when I closed my hand around it.

What is something else that is brittle?

creative

The florist made original and **creative** flower arrangements.

Describe a time when you were creative.

descriptive

The speaker gave a vivid, **descriptive** talk about the rain forest.

Talk about your favorite activity in a descriptive way.

outstretched

The seagull glided through the air on **outstretched** wings.

What is an antonym for outstretched?

Poetry Terms

metaphor

"The garbage truck is a monster" is a **metaphor** because it compares two unlike things.

Give another example of a metaphor.

rhyme

Two words **rhyme** when they sound the same, such as *claw* and *draw*.

What word rhymes with fall?

simile

"The long grass is like hair" is a **simile** because it compares two things using *like* or *as*.

Give another example of a simile.

meter

Meter is the pattern of stressed and unstressed syllables.

How does a strong meter affect the rhythm of a poem?

COLLABORATE

Your Turn

Pick three words. Write three questions for your partner to answer.

Go Digital! *Use the online visual glossary*

DOG

A brown boomerang,
my dog flies off, arcs his way
back into my arms.

— Jeffrey Boyle

? Essential Question

**How are writers
inspired by animals?**

Read how poets use
creative thinking to write
about animals.

THE EAGLE

He clasps the crag with crooked hands;
Close to the sun in lonely lands,
Ring'd with the azure world, he stands.

The wrinkled sea beneath him crawls;
He watches from his mountain walls,
And like a thunderbolt he falls.

— **Alfred, Lord Tennyson**

Alessandra Cimatoribus

151

CHIMPANZEE

From branch to branch on outstretched arms,
From tree to ground I leap.
When I want to eat a snack,
I stick a stick in termite heaps.

I use my teeth to rip off leaves
And make the branch all bare,
Then find the hole the bugs come out
And patiently wait there.

My skinny branch becomes a bridge,
As brittle bugs climb up the stick.
I pick them off one by one
And crunch them like potato chips!

— Ellen Lee

Rat

Teeth like jackhammers,
I chew through concrete for fun,
bring the outdoors in!

— **Rosa Sandoval**

Make Connections

Talk about the creative ways that the poets portray animals. **ESSENTIAL QUESTION**

What animal would you write a poem about? Why? **TEXT TO SELF**

Alessandra Cimatoribus

Lyric Poetry and Haiku

Lyric Poetry:
- Expresses the thoughts and feelings of the poet.
- Often has end rhymes and a consistent meter.

Haiku:
- Uses three short lines to describe a scene or a moment.
- Has a first and last line of five syllables and a second line of seven syllables.

🔍 Find Text Evidence

"The Eagle" is a lyric poem because it tells how the poet feels about the eagle. It also has end rhymes and a consistent meter. "Dog" is a haiku because the lines have the 5-7-5 syllable count.

page 151

THE EAGLE

He clasps the crag with crooked hands;
Close to the sun in lonely lands,
Ring'd with the azure world, he stands.

The wrinkled sea beneath him crawls;
He watches from his mountain walls,
And like a thunderbolt he falls.

— Alfred, Lord Tennyson

151

The poet describes the eagle as "close to the sun in lonely lands." I wonder if the poet feels the eagle is above other creatures in other ways.

Your Turn COLLABORATE

Reread the poem "Rat." Explain what form the poem is and give text evidence.

Point of View

The voice that you hear in a poem is the speaker. The speaker's point of view is how the speaker thinks or feels. Sometimes the speaker is a character in the poem. Sometimes the speaker is telling about the characters or events in the poem.

Find Text Evidence

In "Chimpanzee," the pronouns I *and* me *tell me that the speaker is the chimpanzee. I will reread the poem on page 152 and find the details that give me clues to the Chimpanzee's point of view.*

Details
When I want to eat a snack, I stick a stick in termite heaps.
I use my teeth to rip off leaves.
I pick them off one by one

↓

Point of View
The chimpanzee is confident about finding food.

Your Turn

Reread "The Eagle" on page 151. Is the speaker a character in the poem? List important details that give clues to the speaker's point of view. Then identify the point of view.

Meter and Rhyme

Meter is the rhythm of syllables in a line of poetry. It is created by the arrangement of accented and unaccented syllables. Words **rhyme** when their endings sound the same.

Find Text Evidence

Reread the poem "The Eagle" on page 151. Listen for the end rhymes and to the rhythm of the meter.

page 151

THE EAGLE

He clasps the crag with crooked hands;
Close to the sun in lonely lands,
Ring'd with the azure world, he stands.

The wrinkled sea beneath him crawls;
He watches from his mountain walls,
And like a thunderbolt he falls.

— Alfred, Lord Tennyson

151

Rhyme Say the last words of each line of the first stanza. They rhyme because their endings sound alike.

Meter Read the second stanza aloud. The words are placed to make the syllables seem to bounce. A stressed syllable follows each unstressed syllable.

COLLABORATE

Your Turn

Find other words that rhyme in "The Eagle." Then find out if the meter is the same in every line.

Figurative Language

A **simile** is a comparison made using *like* or *as,* for example, *straight as an arrow*. A **metaphor** is a comparison made without *like* or *as*, for example, *the grass was a green carpet*.

 ### Find Text Evidence

When I read "Chimpanzee" on page 152, I see that the poet uses a simile in the last stanza to describe how the chimpanzee is eating the termites.

I pick them off one by one
And crunch them like potato chips!

 ## Your Turn

 COLLABORATE

Identify the metaphor in "Dog." Then reread "The Eagle" and find a simile. Rewrite the simile as a metaphor.

Alessandra Cimatoribus

Readers to . . .

Poets use precise language by choosing strong verbs and descriptive adjectives to help the reader create a picture in his or her mind. Reread "The Eagle" below.

Precise Language

Identify the **precise language** used in "The Eagle" that makes the poem vivid, interesting, and effective.

Expert Model

He clasps the crag with crooked hands;
Close to the sun in lonely lands,
Ring'd with the azure world, he stands.

The wrinkled sea beneath him crawls;
He watches from his mountain walls,
And like a thunderbolt he falls.

Alessandra Cimatoribus

Writers

Maria wrote a descriptive essay about a raccoon. Read Maria's revisions to one section.

Editing Marks

⌐⌐ Switch order.

∧ Add.

∧ Add a comma.

✐ Take out.

(SP) Check spelling.

≡ Make a capital letter.

/ Make a lowercase letter.

Grammar Handbook

Combining Sentences
See page 457.

Student Model

THE RACCOON

He ~~comes~~ prowls through our yard at

night. He's searching for the fat green plastic

trash cans that he knows will be full

of ~~stuff.~~ delicious treats. He knocks down the cans and

The lids fall off and make a loud noise.

In the beam of my flashlight, his

face looks like a robber's mask.

He freezes for an instant and then

He starts pawing

through the trash

on the driveway.

Your Turn   COLLABORATE

- ✔ Identify the precise language Maria added.
- ✔ Identify the sentences Maria combined.
- ✔ Tell how revisions improved her writing.

Go Digital!
Write online in Writer's Workspace

Unit 3
That's the Spirit!

From
My Country 'Tis of Thee

My country, 'tis of thee,
Sweet land of liberty,
Of thee I sing.
Land where my fathers died,
Land of the pilgrims' pride,
From every mountainside
Let freedom ring!

—Samuel Francis Smith

The Big Idea

How can you show your
community spirit?

Essential Question
How can you make new friends feel welcome?

Go Digital!

A HELPING HAND

Moving to a new place can be intimidating. There are the challenges of going to a new school, making new friends, and learning your way around a new neighborhood. When people are welcoming, all these challenges become a little easier.

▶ What are some things that you can do to help people feel welcome?

▶ What would help you the most on your first day at a new school?

Talk About It

Write words that describe how you would make somebody feel welcome. Then talk with a partner about what somebody has done for you to make you feel welcome.

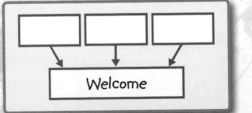

Welcome

Vocabulary

**Use the picture and the sentences to talk with
a partner about each word.**

acquaintance

Joe talked to his new **acquaintance**
Tony in order to get to know him better.

What is the difference between an
acquaintance and a friend?

cautiously

Eric gently and **cautiously** touched the
owl's feathers.

What is a synonym for cautiously?

complementary

Peanut butter and jelly taste good
because they are **complementary** foods.

What two foods do you think are
complementary?

jumble

A **jumble** of masks and snorkels lay
tangled together in the bottom of the boat.

What might you find in a jumble at the
bottom of a closet?

logical

On a multiple-choice test, the girl tried to figure out the most **logical** answers.

What is the most logical way to arrange books in a library?

scornfully

The mother spoke **scornfully** to her son about his bad study habits.

When might you speak scornfully to somebody?

scrounging

Tina saw the cat **scrounging** through the overflowing trash can.

Describe what someone looks like scrounging through a backpack.

trustworthy

When you are mountain climbing, it is good to have a **trustworthy** partner to help you up a cliff.

What is an antonym for trustworthy?

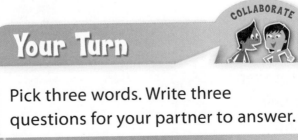

COLLABORATE

Your Turn

Pick three words. Write three questions for your partner to answer.

Go Digital! *Use the online visual glossary*

At the Library

Essential Question

How can you make new friends feel welcome?

Read how a new librarian and an unlikely family become great friends.

166

Rick Dodson admired the pink and orange sky as he waved good-bye to Mrs. Rio and locked the library door. As the sun began its descent behind the Blue Ridge Mountains, Rick started walking to his office to collect his jacket. Seeing a **jumble** of books on a reading table, he sighed and began to gather them into a neat pile.

"No," he stated firmly, and returned the books to the table. "Not tonight."

The librarian never left any books out, but today was his birthday, which meant a brisk walk to the Cupcake Café for a birthday treat before it closed at 5:30 P.M.

That evening, as he sat at home in his book-filled living room, Rick thought about the old friends who had called to wish him a happy birthday. If only this job had not required him to move halfway across the country . . . After six months here, he had made more than one new **acquaintance** but no real friends yet.

"Books are my friends," he thought, which reminded him of the books sitting on the table at the library. "I might as well go back tonight and shelve them," he decided.

He entered the library and flicked on the lights. Immediately, he noticed a book, *Small World*, face down on the floor. "What's going on?" he muttered as he bent down and **cautiously** lifted up the book. "Ahhh," he yelled and dropped the book.

Four miniature figures scrambled out of the way as the book landed on the floor with a thud.

"Mr. Dodson," exclaimed a breathless voice, "we are enchanted to make your acquaintance."

"What . . . who . . . " Rick stammered.

"We're the Bookers! I'm William. This is Emily and our children, Harry and Clementine. By the way, happy birthday!"

"You know it's my birthday?"

"Naturally, we read your file when you arrived six months ago. It's only **logical** that we would want to learn about the new librarian."

"You were **scrounging** through my files?" Rick said, collapsing into a nearby chair. He rubbed his eyes, but the tiny figures were still there—looking up at him expectantly.

Suddenly, the nimble Bookers began shimmying up the table. "We're absolutely **trustworthy**," Emily assured him.

"Haven't you heard of Bookers?" William asked. "Every library has Bookers!"

"We ensure everything runs smoothly," said Emily. "Seen any mice around? They love to gnaw on everything."

Rick slowly shook his head.

"I do nightly rodent patrols," Harry stated proudly. "Those mice run at the sight of me," he added **scornfully**.

"Do your chairs ever squeak?" inquired Clementine. "No! That's because we oil them!"

Rick considered the past six months. He hadn't seen one mouse, his chairs never squeaked, and his pencils were never dull.

"The pencils?" he asked.

"We sharpen them nightly," William replied.

"But why?" asked Rick.

"Look around!" exclaimed William. "We work and read. Bookers and libraries are **complementary**. We belong together."

"To be honest, Mr. Dodson," said Emily, "we wanted to meet you because we thought that we could be friends."

Rick Dodson grinned. "Call me, Rick. And I'd love to be friends," he said.

Rick eventually made other new friends, but he still spent many evenings with the Bookers. He bought a toy car for Harry's rodent patrol, and he read scary stories aloud to Clementine. Every year on his birthday, he brought cupcakes for his friends to share with him.

Make Connections

Talk about how the Bookers made Rick Dodson feel welcome. **ESSENTIAL QUESTION**

How do you make new students in your school feel welcome? **TEXT TO SELF**

Richard Johnson

Visualize

When you read a story, picturing the plot events, characters, and setting in your mind can add to your understanding and enjoyment. As you read the first paragraph of "At the Library," visualize what is happening.

 ## Find Text Evidence

As I read the first paragraph on page 167, I picture Rick Dodson in the library at the end of the day. The descriptive details help me visualize the sunset and Rick waving good-bye to Mrs. Rio.

page 167

Rick Dodson admired the pink and orange sky as he waved good-bye to Mrs. Rio and locked the library door. As the sun began its descent behind the Blue Ridge Mountains, Rick started walking to his office to collect his jacket. Seeing a **jumble** of books on a reading table, he sighed and began to gather them into a neat pile.

"No," he stated firmly, and returned the books to the table. "Not tonight."

The librarian never left any books out, but today was his birthday, which meant a brisk walk to the Cupcake Café for a birthday treat before it closed at 5:30 P.M.

That evening, as he sat at home in his book-filled living room, Rick thought about the old friends who had called to wish him a happy birthday. If only this job had not required him to move halfway across the country . . . After six months here, he had made more than one new **acquaintance** but no real friends yet.

I can visualize Rick looking at the jumble of books on the table and starting to pick them up. I imagine him shaking his head when he decides to wait until the next day.

COLLABORATE

Your Turn

What are some other story events in "At the Library" that you can visualize? As you read, remember to use the strategy Visualize.

Point of View

Every story has a narrator, or a person who tells the story. The narrator's point of view tells how the narrator feels or thinks about the characters or events. When the narrator uses pronouns such as *he, she,* or *they*, the story has a third-person narrator.

Find Text Evidence

When I reread page 167 of "At the Library," I see that the narrator uses the pronouns he, him *and* his *to describe Rick. This story has a third-person narrator. I can use details from the story to figure out the narrator's point of view.*

Details
Rick went out for his birthday treat by himself.
Rick thought about his old friends who had called to wish him a happy birthday.

Pronouns such as *his* and *him* are clues to the narrator.

↓

Point of View
The story has a third-person narrator who understands that Rick misses his old friends and wants new friends.

Your Turn

COLLABORATE

Reread "At the Library." Find other details that give you clues about the narrator's point of view.

Go Digital!
Use the interactive graphic organizer

Fantasy

The selection "At the Library" is a fantasy.

A fantasy:

- Includes invented characters and settings.
- Has elements that could not exist in real life.
- Often has illustrations.

Find Text Evidence

"At the Library" is a fantasy. The story takes place in a library, which is a realistic setting, but the illustrations show that the Bookers are tiny people who would not exist in real life.

page 166

CCSS Shared Read Genre · Fantasy

At the Library

Essential Question
How can you make new friends feel welcome?

Read how a new librarian and an unlikely family become great friends.

166

Use Illustrations Illustrations give the reader visual clues about the story's characters, setting, and events. This illustration shows me what size the Bookers are.

COLLABORATE

Your Turn

Find and list two specific examples in the text that let you know "At the Library" is a fantasy.

172

Context Clues

As you read "At the Library," you may come across words you don't know. To figure out the meaning of an unfamiliar word, look for clues in nearby phrases and sentences.

 Find Text Evidence

When I reread the first paragraph on page 167 of "At the Library," the phrases pink and orange sky *and* behind the Blue Ridge Mountains *help me figure out what* descent *means.*

 Rick Dodson admired the pink and orange sky as he waved good-bye to Mrs. Rio and locked the library door. As the sun began its descent behind the Blue Ridge Mountains, Rick started walking to his office to collect his jacket.

Your Turn

COLLABORATE

Use context clues to find the meanings of the following words in "At the Library." Write an example sentence for each word.

scrambled, *page 167*

gnaw, *page 168*

rodent, *page 168*

Readers to...

Writers use transition words or phrases to make a sequence of events clearer or to move from one idea to another. Reread the excerpt from "At the Library" below.

Transitions

Identify the **transitions**. How do they organize the sequence of events?

Expert Model

That evening, as he sat at home in his book-filled living room, Rick thought about the old friends who had called to wish him a happy birthday. If only this job had not required him to move halfway across the country . . . After six months here, he had made more than one new acquaintance but no real friends yet.

Richard Johnson

174

Writers

Sarah wrote a fantasy. Read Sarah's revisions to one section of her story.

The Old House

Hello!

Once a boy named Jay and his family move^d into an old house. On the first day Jay explored all the rooms. He found a talking owl living in the attic. Later He met a talking groundhog living in the garden. He wanted his two fri⎡e⎦nds to meet. one evening, he held a party in his room. Both animals ~~liked~~ enjoyed the party. After the party Jay ~~got~~ crawled into bed and fell asleep.

Your Turn

COLLABORATE

- ☑ Identify transition words and phrases that Sarah included.
- ☑ Identify an action verb Sarah used.
- ☑ Tell how other revisions improved her writing.

Go Digital!
Write online in Writer's Workspace

175

Essential Question

In what ways can you help your community?

Go Digital!

Planting Hope

People help their communities in different ways. Volunteering at the library, coaching soccer, or turning a vacant lot into a community garden are just some of the ways that people give back to their communities.

► How do people in your community help each other?

► What are some things that you could do to help your community?

Talk About It COLLABORATE

Write words that tell how you can help your community. Then talk to a partner about community projects that you would like to make happen.

Projects

(bkgd) Rudi Von Briel/Photo Edit; (r) maxstock/Alamy

Vocabulary

Use the picture and the sentences to talk with a partner about each word.

assigned

The teacher **assigned** the student extra homework because he was late.

What project has a teacher assigned you recently?

generosity

The man showed his **generosity** by putting twenty dollars in the can for the charity.

What words are associated with generosity?

gingerly

The girl stepped **gingerly** into the waves.

What is a reason why you might step gingerly?

mature

Tom's father said that he was **mature** enough to ride the train by himself.

What is an antonym for mature?

organizations

Students signed up for information about recycling **organizations**.

What are some organizations in your town?

residents

Mrs. Seals enjoyed talking with the **residents** of the nursing home.

What town or city are your classmates residents of?

scattered

The sheep were **scattered** across the meadow.

What is a synonym for scattered?

selective

Tina was **selective** about choosing only the freshest fruits and vegetables.

What are you selective about?

Your Turn

Pick three words. Write three questions for your partner to answer.

Go Digital! *Use the online visual glossary*

REMEMBERING
HURRICANE
KATRINA

Essential Question

In what ways can you help your community?

Read about how Hector helps others after Hurricane Katrina.

Leaning over my steering wheel, I watched the heavy clouds roll in. The sky became a darker shade of gray, and raindrops were soon scattered across my windshield. A storm was coming. Glancing at the boxes of clothes stacked in the backseat, I smiled to myself.

A torrential downpour of rain began beating against my windshield as lightning flickered across the sky. I pulled the car off the road until my driving visibility improved. People on the sidewalk held purses and briefcases over their heads in a futile effort to keep from getting wet. Children screamed and danced around in the downpour. The rain reminded me of another storm ten years earlier.

Hurricane Katrina slammed into the Gulf Coast of the United States when I was nine years old. The ferocious storm caused untold amounts of damage.

One of my strongest memories from that time was watching the evening news with my aunt. A reporter stood inside the Houston Astrodome, surrounded by thousands of people. They all shared the same weary expression. Many wore torn and dirty clothes, and some had no shoes on their feet. They slowly shuffled along, their faces full of sadness.

Tyrone Turner/National Geographic/Getty Images

"Are they here because of the hurricane?" I asked softly.

Aunt Lucia nodded. "*Sí*, Hector. These people are from New Orleans, Louisiana. Just a few days ago, Hurricane Katrina destroyed their homes and possessions, and they lost everything they owned, so now they are temporary **residents** of the Astrodome. It's a place for them to stay until it's safe to go home."

I knew a lot about Katrina. The storm had formed in hot and humid tropical weather and then traveled north. It had come so close to Texas that I worried it would strike us in Houston. It missed us, but other cities were not so lucky.

The TV news reporter looked around. People tried to speak to her, but she was being **selective** about whom she wanted to interview. I noticed a little boy sitting behind her on a cot, hugging an old teddy bear. Watching him, I knew I had to do something.

The next day, my friends joined me at our volunteer club—the Houston Helpers—and together we devised a plan. We wanted to collect toys and give them to the kids at the Astrodome because donating the toys would help bring some happiness into the lives of these families.

Anxious to get started, we made lists of what we needed to do. Then every one of us was **assigned** a specific task.

We agreed to spread the word to our schools and other **organizations**. Three days later, after a Herculean effort on our part, the donation bins were overflowing with new toys!

I'll never forget the day when we entered the Astrodome with our gifts. Children flew toward us from all directions. Smiles lit up their faces as we pulled toys from our bags. Grateful parents thanked us for our **generosity** and complimented our group leaders on how thoughtful and **mature** we all were.

BZZZZ. My cell phone jolted me back to the present, and I noticed that the storm had passed.

"Hector?"

"*Sí*, yes, hi, Jeannie."

"Do you have the donations? A few more families have arrived, more victims of yesterday's tornado."

"Yes, I have the clothing donations. The storm delayed me, but I'll be there soon!"

I **gingerly** eased my car into the suddenly busy traffic. It felt good to know that I was making a difference again.

Make Connections

Talk about how Hector and his friends make a difference in their community.
ESSENTIAL QUESTION

What are some things that you have done to help your school or community? **TEXT TO SELF**

183

(cr) Jeff Manglat

Visualize

When you read fiction, picture the events, characters, and setting in your mind to help you better understand the story. As you read "Remembering Hurricane Katrina," visualize what happens in the story.

Find Text Evidence

In the first two paragraphs of "Remembering Hurricane Katrina" on page 181, I can use the details to picture the setting. The narrator describes the rain, the lightning, and the people on the sidewalk holding briefcases and purses over their heads.

page 181

Leaning over my steering wheel, I watched the heavy clouds roll in. The sky became a darker shade of gray, and raindrops were soon scattered across my windshield. A storm was coming. Glancing at the boxes of clothes stacked in the backseat, I smiled to myself.

A torrential downpour of rain began beating against my windshield as lightning flickered across the sky. I pulled the car off the road until my driving visibility improved. People on the sidewalk held purses and briefcases over their heads in a futile effort to keep from getting wet. Children screamed and danced around in the downpour. The rain reminded me of another storm ten years earlier.

Hurricane Katrina slammed into the Gulf Coast of the United States when I was nine years old. The ferocious storm caused untold amounts of damage.

One of my strongest memories from that time was watching the evening news with my aunt. A reporter

I can use these descriptive details to visualize what the setting looks and sounds like.

Your Turn

COLLABORATE

Visualize the scene between Hector and his aunt as they watch the news report. Describe what you "see" to a partner. As you read, remember to use the strategy Visualize.

Point of View

The narrator's point of view tells how the narrator thinks or feels about characters or events in the story. When the narrator uses the pronouns *I, me,* or *my,* the story is told by a first-person narrator.

 Find Text Evidence

When I reread page 151 of "Remembering Hurricane Katrina," I see that the narrator uses the pronouns I, me, *and* my. *That tells me the story is told by a first-person narrator, Hector. I can find clues in the text about the narrator's point of view.*

Details
Hector remembers watching the hurricane victims slowly shuffling along with faces full of sadness.
Hector noticed a little boy hugging an old teddy bear and realized he had to do something.

↓

Point of View
The narrator, Hector, thinks it is important to help the hurricane victims.

Your Turn

Reread "Remembering Hurricane Katrina." Find other details that tell the point of view.

Go Digital!
Use the interactive graphic organizer

185

Realistic Fiction

The selection "Remembering Hurricane Katrina" is realistic fiction.

Realistic fiction:
- Is a made-up story.
- Includes realistic characters, events, and settings.
- Usually has dialogue.
- May include a flashback to an earlier event.

Find Text Evidence

I can tell "Remembering Hurricane Katrina" is realistic fiction.
The characters, events, and setting could all exist in real life.
The story has dialogue and includes a flashback.

page 181

Leaning over my steering wheel, I watched the heavy clouds roll in. The sky became a darker shade of gray, and raindrops were soon scattered across my windshield. A storm was coming. Glancing at the boxes of clothes stacked in the backseat, I smiled to myself.

A torrential downpour of rain began beating against my windshield as lightning flickered across the sky. I pulled the car off the road until my driving visibility improved. People on the sidewalk held purses and briefcases over their heads in a futile effort to keep from getting wet. Children screamed and danced around in the downpour. The rain reminded me of another storm ten years earlier.

Hurricane Katrina slammed into the Gulf Coast of the United States when I was nine years old. The ferocious storm caused untold amounts of damage.

One of my strongest memories from that time was watching the evening news with my aunt. A reporter stood inside the Houston Astrodome, surrounded by thousands of people. They all shared the same weary expression. Many wore torn and dirty clothes, and some had no shoes on their feet. They slowly shuffled along, their faces full of sadness.

Flashback Sometimes authors do not present a story's events in time order. Authors might take the reader back to an event that happened in the past. This is called a *flashback*.

Your Turn

COLLABORATE

Find and list two more examples in "Remembering Hurricane Katrina" that show it is realistic fiction.

Context Clues

As you read "Remembering Hurricane Katrina," you may come across a word that you don't know. A definition of the word may be in the text nearby, or the word may be restated in a simpler way. Sometimes an example may be given. You can use these context clues to figure out the word's meaning.

Find Text Evidence

When I read the fifth paragraph on page 182 of "Remembering Hurricane Katrina," the phrase collect toys and give them *helps me figure out what the word* donating *means.*

We wanted to collect toys and give them to the kids at the Astrodome because donating the toys would help bring some happiness into the lives of these families.

COLLABORATE

Your Turn

Use context clues to figure out the meanings of the following words in "Remembering Hurricane Katrina."

shuffled, *page 181*
possessions, *page 182*
delayed, *page 183*

Readers to...

Writers use strong, concrete words and sensory details to describe the action in a story. Reread an excerpt from the beginning of "Remembering Hurricane Katrina" below.

Expert Model

Strong Words

Identify **strong words** used in the paragraph. How do these words make the story more interesting to read?

A torrential downpour of rain began beating against my windshield as lightning flickered across the sky. I pulled the car off the road until my driving visibility improved. People on the sidewalk held purses and briefcases over their heads in a futile effort to keep from getting wet. Children screamed and danced around in the downpour. The rain reminded me of another storm ten years earlier.

Writers

Editing Marks

⌐⌐ Switch order.

∧ Add.

∧ Add a comma.
  ͵

ℒ Take out.

SP Check spelling.

≡ Make a capital letter.

Grammar Handbook

Verb Tenses See page 458.

Reina wrote a story. Read Reina's revisions to one section of her story.

Student Model

Harriet and Me

Standing in the ~~darkness in the~~ *shadow of the moon*

~~night,~~ I paced back and forth. I was

hidden by the forest, but I ~~fear~~ *feared* that

secret eyes were ~~out there~~ *hunting me*. Harriet

stood nearby. It was my first time

helping her transport ∧ on the *passengers*

Underground Railroad. I was ~~worry~~ *worried*

that I would fail her. I heard a loud

~~noise~~ *rustling and crackling* in the bushes, and I felt my

whole body freeze in terror.

Your Turn

COLLABORATE

☑ Identify strong words that Reina included.

☑ Identify the verb tense that she corrected in her writing.

☑ Tell how Reina's revisions improved her writing.

Go Digital!
Write online in Writer's Workspace

189

Weekly Concept Liberty and Justice

Essential Question

How can one person make a difference?

Go Digital!

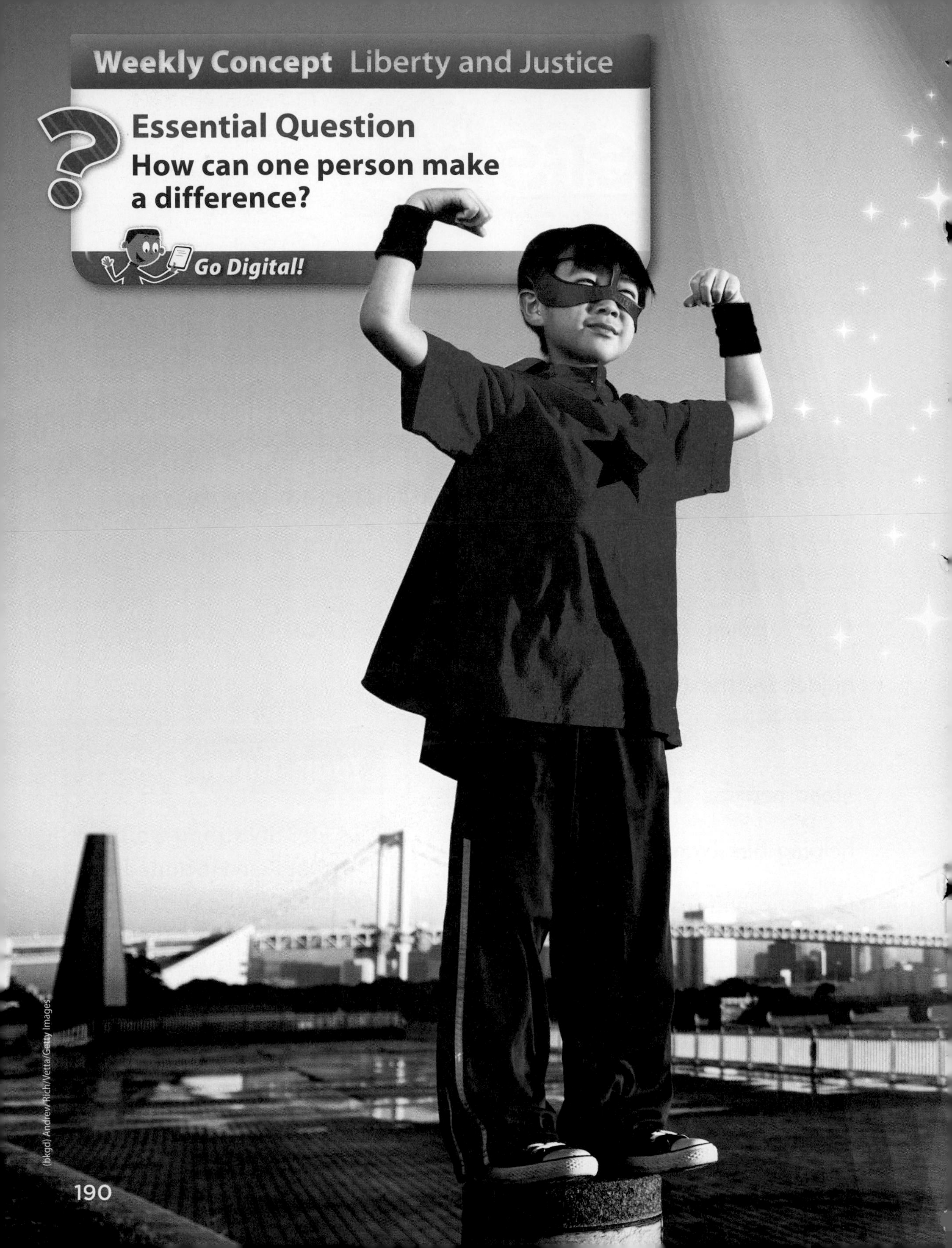

EVERYDAY HEROES

Comic book superheroes such as Superman and Batman are famous for protecting people. However, in real life, it is the everyday heroes—people who speak out against injustice and inequality, people who work to help others—who are the real superheroes.

▶ How has somebody made a difference in your life?

▶ Do you think it is possible for one person to create change and make a difference?

▶ What could you do to make a difference?

Talk About It

Write words that describe how people can make a difference. Then talk to a partner about what you could do to make a difference.

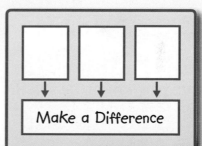

Make a Difference

Vocabulary

Use the picture and the sentences to talk with a partner about each word.

boycott

Joan bought apples instead of grapes after she joined the grape **boycott.**

Why might a boycott help change laws?

encouragement

The **encouragement** we needed to win the game came from our fans.

What kind of encouragement do you give others?

fulfill

Jules got to **fulfill** his dream of performing in the school talent show.

What dream would you like to fulfill one day?

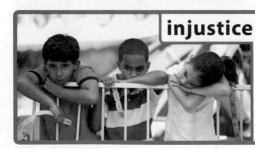

injustice

The children felt that it was an **injustice** that they were not allowed on the roller coaster because they were too short.

How are injustice and justice related?

mistreated

Tom felt that the dog's former owner had harmed and **mistreated** her.

What is an antonym for mistreated?

protest

The children decided to **protest** the destruction of the forest.

What is a synonym for protest?

qualified

Doctor Smith is more **qualified** than the nurse to tell what treatment the boy needs.

What would you need to do to be qualified to teach math?

registered

The woman gave her address so that she could be **registered** to vote.

Why is it be important to be registered to vote?

Your Turn

COLLABORATE

Pick three words. Write three questions for your partner to answer.

Go Digital! *Use the online visual glossary*

Judy's APPALACHIA

A mountaintop is leveled to mine for coal in Appalachia.

? **Essential Question**

How can one person make a difference?

Read about how one person decided to take a stand.

Judy Bonds's six-year-old grandson stood in a creek in West Virginia. He held up a handful of dead fish and asked, "What's wrong with these fish?" All around him dead fish floated belly up in the water. That day became a turning point for Judy Bonds. She decided to fight back against the coal mining companies that were poisoning her home.

Marfork, West Virginia

The daughter of a coal miner, Julia "Judy" Bonds was born in Marfork, West Virginia in 1952. The people of Marfork had been coal miners for generations because coal mining provided people with jobs. Coal gave people the energy they needed to light and warm their homes.

But Marfork wasn't just a place where coal miners lived. Marfork was home to a leafy green valley, or holler, surrounded by the Appalachian Mountains on every side. Judy's family had lived in Marfork for generations. Judy grew up there swimming and fishing in the river. She raised a daughter there.

Mountaintop Removal Mining

An energy company came to Marfork in the 1990s. It began a process called mountaintop removal mining. Using dynamite, the company blew off the tops of mountains to get at the large amounts of coal underneath. The process was quicker than the old method of digging for coal underground, but it caused many problems. Whole forests were destroyed.

Judy Bonds spoke out against mountaintop removal mining.

Dust from the explosions filled the air and settled over the towns. Coal sludge, a mixture of mud, chemicals, and coal dust, got into the creeks and rivers.

Pollution from the mountaintop removal mining began making people living in the towns below the mountains sick. In the area where Judy lived, coal sludge flowed into the rivers and streams. People packed up and left. Judy was heartbroken. The land she loved was being **mistreated**. She realized that the valley that had always been her home had been poisoned. No longer a safe place to live, it had become dangerous. Judy, her daughter, and her grandson had to leave.

Working for Change

Something had to be done about the pollution. Judy decided it was important to **protest** against strip mining and demand that it be stopped. She felt that she must try to keep the area safe for people. She felt **qualified** to talk to groups about the **injustice** of whole towns being forced to move and mountains and forests being destroyed, all because of strip mining. After all, she had grown up in a mining family.

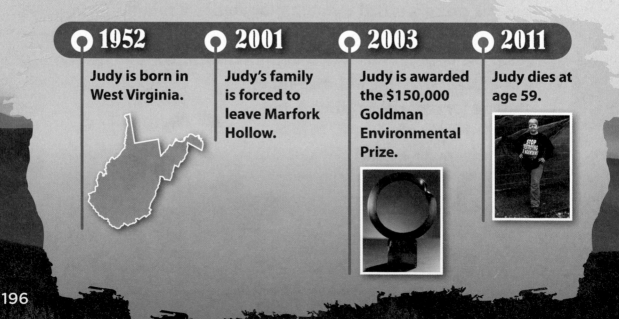

1952	**2001**	**2003**	**2011**
Judy is born in West Virginia.	Judy's family is forced to leave Marfork Hollow.	Judy is awarded the $150,000 Goldman Environmental Prize.	Judy dies at age 59.

Judy worked as a volunteer for the Coal River Mountain Watch, a group that fought against mountaintop removal mining. Eventually, she became its executive director. She **registered** to take part in protests against mining companies. At the protests, Judy faced a lot of anger and insults. Many coal miners were not opposed to mountaintop removal mining. They supported it because they needed the jobs to provide for their families. Judy knew it would be impossible to **boycott** the mining companies. The coal miners could not afford to leave their jobs. Instead, she pushed for changes to be made to the mining process. Slowly, small changes were made to protect communities in mining areas. In 2003, Judy was awarded the Goldman Environmental Prize for her efforts as an activist.

Judy Bonds spoke at protests.

Remembering Judy

Sadly, Judy could not **fulfill** all of her goals. She was diagnosed with cancer and died in January 2011. But her success has provided **encouragement** to other activists. Judy may not have been able to stay in her home, but her work will help preserve and protect the Appalachian Mountains and help others remain in their homes.

The Monongahela National Forest in West Virginia

Make Connections

How did Judy Bonds make a difference? **ESSENTIAL QUESTION**

What cause do you feel strongly about? **TEXT TO SELF**

Reread

When you read an informational text, you may come across information and facts that are new to you. As you read "Judy's Appalachia," reread sections of text to make sure you understand and remember the information.

🔍 Find Text Evidence

You may not be sure what mountaintop removal mining is. Reread page 195 of "Judy's Appalachia."

page 195

surrounded by the Appalachian Mountains on every side. Judy's family had lived in Marfork for generations. Judy grew up there swimming and fishing in the river. She raised a daughter there.

Mountaintop Removal Mining

An energy company came to Marfork in the 1990s. It began a process called mountaintop removal mining. Using dynamite, the company blew off the tops of mountains to get at the large amounts of coal underneath. The process was quicker than the old method of digging for coal underground, but it caused many problems. Whole forests were destroyed.

Judy Bonds spoke out against mountaintop removal mining.

I read that this kind of mining is a way of getting coal by blowing off the top of a mountain to get to the coal underneath.

COLLABORATE

Your Turn

Why did Judy Bonds leave Marfork? Reread page 196 of "Judy's Appalachia" to answer the question. As you read, remember to use the strategy Reread.

Author's Point of View

Authors have a position or point of view about the topics they write about. Look for details in the text, such as the reasons and evidence the author chooses to present. This will help you to figure out the author's point of view.

 Find Text Evidence

When I reread the top of page 195, I can look for details that reveal the author's point of view about Judy Bonds.

Details
Judy sees her grandson in a creek surrounded by dead fish.
Judy decides to fight the mining companies. They are poisoning her home.

Look at the evidence the author presents.

↓

Author's Point of View
The author admires Judy Bonds for taking a stand against the coal mining companies.

COLLABORATE

Your Turn

Reread pages 196–197. Look for two more details that help support the author's point of view and list them in your graphic organizer.

Go Digital!
Use the interactive graphic organizer

199

Biography

The selection "Judy's Appalachia" is a biography.

Biography:

- Is the story of a real person's life written by another person.
- Usually presents events in chronological order.
- May include text features.

Find Text Evidence

"Judy's Appalachia" is a biography. The text describes a real person. The events in Judy's life are presented in chronological order. There are text features.

page 196

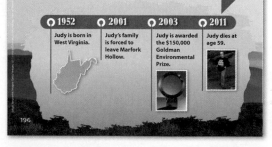

Dust from the explosions filled the air and settled over the towns. Coal sludge, a mixture of mud, chemicals, and coal dust, got into the creeks and rivers.

Pollution from the mountaintop removal mining began making people living in the towns below the mountains sick. In the area where Judy lived, coal sludge flowed into the rivers and streams. People packed up and left. Judy was heartbroken. The land she loved was being **mistreated**. She realized that the valley that had always been her home had been poisoned. No longer a safe place to live, it had become dangerous. Judy, her daughter, and her grandson had to leave.

Working for Change

Something had to be done about the pollution. Judy decided it was important to **protest** against strip mining and demand that it be stopped. She felt that she must try to keep the area safe for people. She felt **qualified** to talk to groups about the **injustice** of whole towns being forced to move and mountains and forests being destroyed, all because of strip mining. After all, she had grown up in a mining family.

1952 — Judy is born in West Virginia. | 2001 — Judy's family is forced to leave Marfork Hollow. | 2003 — Judy is awarded the $150,000 Goldman Environmental Prize. | 2011 — Judy dies at age 59.

196

Text Feature

Time line A time line is a kind of diagram that shows events in the order in which they took place.

Your Turn

COLLABORATE

Find and list two text features in "Judy's Appalachia." Tell your partner what information you learned from each of the features.

Synonyms and Antonyms

As you read "Judy's Appalachia," you may come across an unfamiliar word. Sometimes the author will use a synonym or an antonym that will help you figure out the meaning of the word. Synonyms are words with similar meanings. Antonyms are words with opposite meanings.

Find Text Evidence

When I read page 197 in "Judy's Appalachia," I do not know what the word opposed *means. The word* supported *is in the next sentence. I know that* supported *means "in favor of." This will help me figure out what* opposed *means.*

> Many coal miners were not opposed to mountaintop removal mining. They supported it because they needed the jobs to provide for their families.

Your Turn

With a partner, use synonyms or antonyms to find the meanings of the following words.

method, *page 195*
dangerous, *page 196*
preserve, *page 197*

Panoramic Images/Getty Images

Readers to...

Writers include reasons and evidence to support their opinions about a topic. They provide facts, details, and examples, to help prove their point of view. Reread the paragraph from "Judy's Appalachia" below.

Expert Model

Relevant Evidence

Identify the author's opinion of mining. What **reasons** and **evidence** does the author use to support that opinion?

Pollution from the mountaintop removal mining began making people living in the towns below the mountains sick. In the area where Judy lived, coal sludge flowed into the rivers and streams. People packed up and left. Judy was heartbroken. The land she loved was being mistreated. She realized that the valley that had always been her home had been poisoned. No longer a safe place to live, it had become dangerous. Judy, her daughter, and her grandson had to leave.

Writers

Max wrote about his Uncle Ryan. Read Max's revisions to one section of his essay.

Uncle Ryan

~~I think~~ my Uncle Ryan made a big

difference in our neighborhood. He

started a recycling program and got

People now volunteer to pick up old
newspapers, cans, and bottles.

many people involved. Before the ^

program, our neighborhood had too

much trash on the streets. The local

was filling
landfill ~~were fill~~ up too fast. Lots of ^

people were complaining. Uncle Ryan

did a lot of work to get the recycling

He went to town meetings for months!
program working. ^

Editing Marks

⌐⌐ Switch order.

∧ Add.

⌃ Add a comma.

ℐ Take out.

ⓢⓟ Check spelling.

≡ Make a capital letter.

Grammar Handbook

Main and Helping Verbs See page 460.

Your Turn COLLABORATE

☑ Identify the reasons for Max's opinion.
☑ Identify a helping verb.
☑ Tell how revisions improved his writing.

Go Digital!
Write online in Writer's Workspace

Essential Question
How can words lead to change?

Go Digital!

"Education is the most powerful weapon which you can use to change the world."

—NELSON MANDELA

(bkgd) SZ Photo/The Bridgeman Art Library

204

Listen & Learn

The man in the photograph is Nelson Mandela. He is a famous activist and statesman who fought a long battle for equality in South Africa. His words have inspired people all over the world to fight against racism and injustice.

► Read Mandela's words on page 204. What do you think he means?

► How could your words lead to change?

Talk About It COLLABORATE

Write phrases that describe how people use powerful words to create change. Then talk to a partner about Mandela's words.

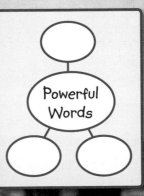

Powerful Words

Vocabulary

**Use the picture and the sentences to talk with
a partner about each word.**

address

In his **address**, the mayor urged the
citizens to take pride in their community.

What is a synonym for address?

divided

The three interviewers were **divided**
about hiring the woman.

Describe a time you felt divided
about something.

haste

Milo had to make **haste** in order not to
be late for class.

What is a antonym for haste?

opposed

The catcher was **opposed** to the
umpire's call and wanted to protest it.

Why might a parent be opposed to
having a pet?

perish

If you put the lettuce in the refrigerator, it will stay fresh and not **perish** as quickly.

What is a synonym for perish?

proclamation

The Town Crier rang her bell before she read the **proclamation** from the mayor.

How are the words proclaim and proclamation related?

shattered

There was lightning, and then a thunderclap **shattered** the silence of the night.

What other things can be shattered?

tension

Tony felt **tension** as he tried to remember what the correct answer was on the computer test.

What is an antonym for tension?

Your Turn

COLLABORATE

Pick three words. Write three questions for your partner to answer.

Go Digital! *Use the online visual glossary*

Words for Change

Essential Question

How can words lead to change?

Read how Elizabeth Cady Stanton's words helped bring about change for women.

The Early Years

In 1827, when Elizabeth Cady Stanton was eleven, her father said: "Oh, my daughter, I wish you were a boy." Elizabeth was **shattered**. From that time on, she became determined to prove to her father and the whole world that women—and all people—deserve equal treatment.

Elizabeth's father was a lawyer, judge, and congressman. She would listen eagerly when a woman would come see him for legal advice. But she was often disappointed. Her father could not help them because women did not have the same rights as men did under the law. Married women could not own property or vote. Elizabeth said: "The tears and complaints of the women who came to my father for legal advice touched my heart and early drew my attention to the injustice and cruelty of the laws."

Elizabeth began drawing lines through all the laws she **opposed** in her father's law books. She planned to take a pair of scissors and cut these pages out. Her father had a better idea. He told her that when she was grown up, she must get lawmakers to pass new laws. Then the unfair laws would **perish** and disappear. Women's lives would be changed.

FULL SUFFRAGE FOR WOMEN

Suffragettes march in a parade in New York City.

209

Working for Change

Elizabeth was as passionate about the rights of African Americans as she was about those of women. At that time, the country was **divided** in two by the issue of slavery. While working for reform, she met her husband, the abolitionist Henry Stanton. They were married in 1840. Elizabeth refused to use the traditional words "promise to obey" in her wedding vows.

The Seneca Falls Convention

Elizabeth tried to settle into the role of wife and mother. But she wanted to be an activist and work for change. She took her father's advice and wrote a **proclamation**. It was called the "Declaration of Rights and Sentiments." Modeled after the Declaration of Independence, it stated that women should be able to vote and have the same rights as men.

She presented this document in 1848 at America's first women's rights convention in Seneca Falls, New York. Elizabeth and her friend Lucretia Mott organized this important event. In her **address** at the convention, Elizabeth said,

Because women do feel themselves...deprived of their most sacred rights, we insist that they have immediate admission to all the rights and privileges which belong to them as citizens of the United States.

List of attendees at the Convention

Our Roll of Honor
Containing all the Signatures to the "Declaration of Sentiments" Set Forth by the First
Woman's Rights Convention,
held at
Seneca Falls, New York
July 19-20, 1848

Lucretia Mott
Harriet Cady Eaton
Margaret Pryor
Elizabeth Cady Stanton
Eunice Newton Foote
Mary Ann M'Clintock
Margaret Schooley
Martha C. Wright
Jane C. Hunt
Amy Post
Catherine F. Stebbins
Mary Ann Frink
Lydia Mount
Delia Mathews
Catherine C. Paine
Elizabeth W. M'Clintock
Malvina Seymour
Phebe Mosher
Catherine Shaw
Deborah Scott
Sarah Hallowell
Mary M'Clintock
Mary Gilbert

LADIES:

Sophronia Taylor
Cynthia Davis
Hannah Plant
Lucy Jones
Sarah Whitney
Mary H. Hallowell
Elizabeth Conklin
Sally Pitcher
Mary Conklin
Susan Quinn
Mary S. Mirror
Phebe King
Julia Ann Drake
Charlotte Woodward
Martha Underhill
Dorothy Mathews
Eunice Barker
Sarah R. Woods
Lydia Gild
Sarah Hoffman
Elizabeth Leslie
Martha Ridley

Rachel D. Bonnel
Betsey Tewksbury
Rhoda Palmer
Margaret Jenkins
Cynthia Fuller
Mary Martin
P. A. Culvert
Susan R. Doty
Rebecca Race
Sarah A. Mosher
Mary E. Vail
Lucy Spalding
Lovina Latham
Sarah Smith
Eliza Martin
Maria E. Wilbur
Elizabeth D. Smith
Caroline Barker
Ann Porter
Experience Gibbs
Antoinette E. Segur
Hannah J. Latham
Sarah Sisson

Richard P. Hunt
Samuel D. Tillman
Justin Williams
Elisha Foote
Frederick Douglass
Henry W. Seymour
Henry Seymour
David Spalding
William G. Barker
Elias J. Doty
John Jones

GENTLEMEN:

William S. Dell
James Mott
William Burroughs
Robert Smallbridge
Jacob Mathews
Charles L. Hoskins
Thomas M'Clintock
Saron Phillips
Jacob P. Chamberlain
Jonathan Metcalf

Nathan J. Milliken
S. E. Woodworth
Edward F. Underhill
George W. Pryor
Joel Bunker
Isaac Van Tassel
Thomas Dell
E. W. Capron
Stephen Shear
Henry Hatley

VOTES FOR WOMEN

VOTES FOR WOMEN
WOMEN'S POLITICAL UNION

A Winning Team

Three years later, Elizabeth met Susan B. Anthony. Together, the two made an unstoppable team. Elizabeth was a passionate speaker and writer. Anthony was a gifted leader and organizer. In 1869, they formed the National Woman Suffrage Association. This group was dedicated to helping women gain the right to vote. Congress showed no **haste**, or hurry, to change the law. Elizabeth toured the country. She spoke about reforms for women and a woman's right to vote. She did not care if her speeches caused **tension** and made some people angry. She believed in her cause.

Victory At Last

Elizabeth Cady Stanton never got to cast a vote before she died on October 26, 1902. Yet her bold words had a lasting impact. Women finally gained the right to vote on August 18, 1920 when the 19th amendment was ratified. Elizabeth Cady Stanton's passion for equal rights paved the way for future women's lives to be changed forever.

Make Connections

Talk about how Elizabeth Cady Stanton helped women gain the right to vote. **ESSENTIAL QUESTION**

Think about a time when you disagreed with something or wanted to change something. What did you say to try to change it? **TEXT TO SELF**

Reread

When you read an informational text, you will often come across new facts and ideas that you would like to remember. As you read "Words for Change," stop and reread key sections to help you understand and remember the information.

 Find Text Evidence

You may not be sure how the Seneca Falls Convention came about. Reread page 210 of "Words for Change" to find out.

page 210

The Seneca Falls Convention

Elizabeth tried to settle into the role of wife and mother. But she wanted to be an activist and work for change. She took her father's advice and wrote a **proclamation**. It was called the "Declaration of Rights and Sentiments." Modeled after the Declaration of Independence, it stated that women should be able to vote and have the same rights as men.

She presented this document in 1848 at America's first women's rights convention in Seneca Falls, New York. Elizabeth and her friend Lucretia Mott organized this important event. In her **address** at the convention, Elizabeth said,

Because women do feel themselves...deprived of their most sacred rights, we insist that they have immediate admission to all

Our Roll of Honor

Signatures to the "Declaration of Sentiments" Set Forth in the First

Woman's Rights Convention,

Seneca Falls, New York
July 19–20, 1848

When I reread, I learn that Elizabeth and Lucretia organized this important event. From this text evidence, I can infer that Elizabeth and Lucretia were determined women who got things done.

Your Turn

COLLABORATE

Why were Elizabeth Cady Stanton and Susan B. Anthony a good team? Reread the section "A Winning Team" on page 211 and answer the question. As you read, remember to use the strategy Reread.

Author's Point of View

The author's point of view is his or her position or attitude about the topic of the selection. Looking closely at the reasons and evidence presented in the text will help you figure out how the author feels about the topic.

 Find Text Evidence

When I reread page 210 of "Words for Change," I can look for details that show how the author feels about Elizabeth Cady Stanton and her fight for women's rights.

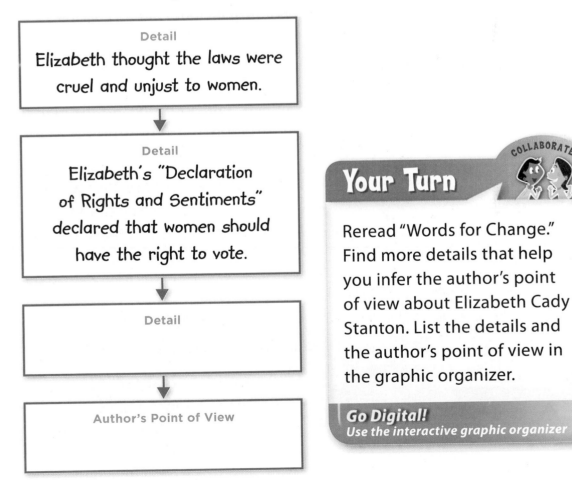

Detail

Elizabeth thought the laws were cruel and unjust to women.

Detail

Elizabeth's "Declaration of Rights and Sentiments" declared that women should have the right to vote.

Detail

Author's Point of View

Your Turn

Reread "Words for Change." Find more details that help you infer the author's point of view about Elizabeth Cady Stanton. List the details and the author's point of view in the graphic organizer.

Go Digital!
Use the interactive graphic organizer

213

Biography

"Words for Change" is a biography.

A biography:
- Is the story of a real person's life written by another person.
- Usually presents events in chronological order.
- May include text features such as primary sources.

Find Text Evidence

"Words for Change" is a biography. It tells about an important actual person, Elizabeth Cady Stanton. It includes primary sources.

page 210

Working for Change

Elizabeth was as passionate about the rights of African Americans as she was about those of women. At that time, the country was **divided** in two by the issue of slavery. While working for reform, she met her husband, the abolitionist Henry Stanton. They were married in 1840. Elizabeth refused to use the traditional words "promise to obey" in her wedding vows.

The Seneca Falls Convention

Elizabeth tried to settle into the role of wife and mother. But she wanted to be an activist and work for change. She took her father's advice and wrote a **proclamation**. It was called the "Declaration of Rights and Sentiments." Modeled after the Declaration of Independence, it stated that women should be able to vote and have the same rights as men.

She presented this document in 1848 at America's first women's rights convention in Seneca Falls, New York. Elizabeth and her friend Lucretia Mott organized this important event. In her **address** at the convention, Elizabeth said,

Because women do feel themselves...deprived of their most sacred rights, we insist that they have immediate admission to all the rights and privileges which belong to them as citizens of the United States.

List of attendees at the Convention

210

Text Feature

Captions Captions give additional information about photographs and text features.

Primary Source Primary sources are original works such as a diaries, letters, or documents created at the time of the event.

COLLABORATE

Your Turn

Find another primary source in "Words for Change." Tell your partner what you learned from it.

Latin and Greek Suffixes

A suffix is a word part added to the end of a word to change its meaning. Some suffixes come from Latin, such as:

-ment = the act or process of
-able = capable of

Other suffixes come from Greek, such as:
-ist = one who has a profession

Find Text Evidence

As I read page 209 of "Words for Change" I am not sure what treatment *means. I know the base word* treat *means behave towards. The suffix* -ment *means the act or process of.*

From that time on, she became determined to prove to her father and the whole world that women—and all people—deserve equal ⌐treatment.¬

COLLABORATE

Your Turn

Use your knowledge of suffixes to find the meanings of the following words in "Words for Change."

activist, *page 210*
unstoppable, *page 211*
amendment, *page 211*

Readers to . . .

An essay or article often ends with a strong conclusion that sums up the author's ideas or opinion. Reread the concluding paragraph from "Words for Change" below.

Expert Model

Strong Conclusions

Identify the **concluding statement** that sums up the author's opinion of Elizabeth Cady Stanton.

Elizabeth Cady Stanton never got to cast a vote before she died on October 26, 1902. Yet her bold words had a lasting impact. Women finally gained the right to vote on August 18, 1920 when the 19th amendment was ratified. Elizabeth Cady Stanton's passion for equal rights paved the way for future women's lives to be changed forever.

Writers

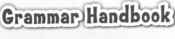

Maya wrote a personal narrative. Read Maya's revisions to the conclusion of her narrative.

Editing Marks

⊓ Switch order.

∧ Add.

⌄ Add a comma.

♪ Take out.

(SP) Check spelling.

≡ Make a capital letter.

Grammar Handbook

Linking Verbs See page 461.

Student Model

The True Story

When Anna

∧ ~~After my friend had~~ explained

to me why she did not come to my

had ♪
party, I realized that I ~~have~~ jumped
∧

to conclusions. It was not fair of me

to judge ⎡without⎤ ⎡her⎤ listening to her

experience ♪
side of the story. This whole ~~thing~~
∧

has taught me an important lesson.

From now on ⌄ I will make sure that I

have
~~has~~ all the information before I draw
∧

a conclusion about a situration. (SP)

Your Turn

☑ Identify Maya's concluding statement.
☑ Identify a linking verb in her essay.
☑ Tell how revisions improved Maya's writing.

Go Digital!
Write online in Writer's Workspace.

217

Essential Question
In what ways can advances in science be helpful or harmful?

Go Digital!

F🍎🍎D for THOUGHT

Scientific discoveries have led to disease-resistant types of corn, rice, and other crops. Some people believe these discoveries will help solve world hunger. Others believe these new foods will cause more harm than good.

▶ Why might it be a good idea for scientists to help farmers grow healthier crops?

▶ Why might it be a bad idea for scientists to interfere with nature?

Talk About It

Write words that describe the pros and cons of advances in science. Then talk with a partner about what your opinion is on the topic.

Scientific Advances

Vocabulary

Use the picture and the sentences to talk with a partner about each word.

advancements

New **advancements** in technology have made satellite dishes more efficient.

What are some examples of scientific advancements?

agriculture

The farmer studied **agriculture** so he could grow healthier crops.

What else might you learn from studying agriculture?

characteristics

Thorns and brightly colored petals are two **characteristics** of a rose.

What characteristics does a cat have?

concerns

The doctor shared his **concerns** about his patient's health.

What concerns might you have if you went to a new school?

disagreed

The two girls **disagreed** about whose turn it was to choose a game to play.

What is an antonym for disagreed?

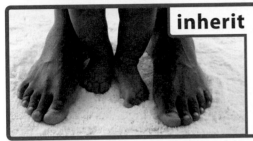

inherit

Shane hopes his baby will not **inherit** his big feet.

What other characteristics can we inherit from our parents?

prevalent

Snowstorms are widespread and **prevalent** in the North.

What plants are prevalent in your neighborhood?

resistance

Drinking lots of water and exercising builds an athlete's **resistance** to illness.

What might strengthen your resistance to illness during the winter?

Your Turn

COLLABORATE

Pick three words. Write three questions for your partner to answer.

Go Digital! **Use the online visual glossary**

(bkgd) John Lamb/Stone/Getty Images

Essential Question

In what ways can advances in science be helpful or harmful?

Read about how science has helped to make better food crops.

222

Food Fight

Is it safe to interfere with Mother Nature?

An incredible thing is happening to our food. Some scientists are using a technique called genetic modification to make superior food crops. It involves altering a seed's genes. Genes are the "instruction codes" that all living things have inside their cells. A seed's genetic code sets what **characteristics** it will **inherit** when it grows into a plant. These could mean how big it will grow and the nutrients it will contain.

For thousands of years, farmers made crops better by crossbreeding plants. They would add pollen from the sweetest melon plants to the flowers of plants that produced the biggest melons. This would make new plants with big, sweet melons. But this process does not always work. The cycle of crossbreeding can take years to get good results.

But advances in gene science have created amazing shortcuts. Using new tools, scientists can put a gene from one living thing into another.

That living thing could be a plant, a bacterium, a virus, or even an animal. These foods are called genetically modified foods, or GM foods. The goal of GM foods is to create foods that can survive insects or harsh conditions or can grow faster. But are these **advancements** in **agriculture** good for us?

(t) Annabelle Breakey/Digital Vision/Getty Images

Support for Superfoods

Scientists believe the new techniques can create crops with a **resistance** to pests and disease. Bt corn is a genetically modified corn.

It has an insect-killing gene that comes from a bacterium. Farmers who grow Bt corn can use fewer chemicals while they grow their crops. That is good for the farmer and the environment.

Some superfoods are extra nutritious. Golden rice has been

Disease-resistant GM potatoes were introduced in the 1990's.

genetically modified with three different genes. One gene is a form of bacterium. The other two are from daffodils. The new genes help the rice to make a nutrient that prevents some forms of blindness.

Superfoods

These foods may seem common. But did you know that the genetically modified versions have special powers?

Rice

Rice contains phytic acid. Too much of this acid can be bad for people. A new type of rice has been bred with a low level of phytic acid.

Salmon

To create supersized salmon, scientists changed the gene that controls growth. The genetically altered salmon grow twice as fast as their wild cousins.

Tomatoes

Genetically engineered tomatoes can be picked when they are ripe and still not bruise when shipped. One food company tried to use an arctic flounder fish gene to create a tomato that could survive frost. The fish-tomato did not succeed.

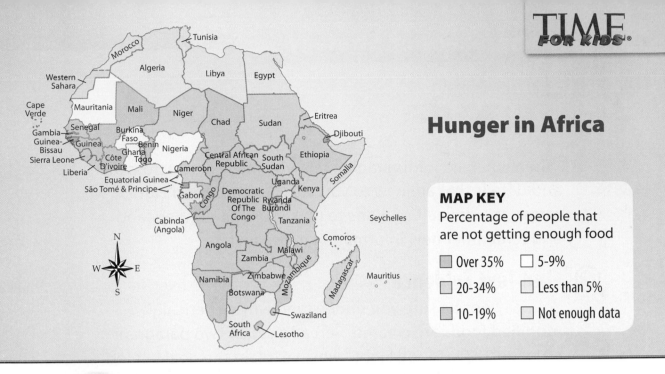

Hunger in Africa

MAP KEY

Percentage of people that are not getting enough food

☐ Over 35%	☐ 5-9%
☐ 20-34%	☐ Less than 5%
☐ 10-19%	☐ Not enough data

Safety Issues

Many people have **disagreed** with the idea that GM foods are a good idea. They worry GM foods will hurt the environment and humans. One concern is that plants with new genes will crossbreed with weeds to make pesticide-resistant weeds. Another concern is that GM foods may trigger allergies.

Genetically modified crops are **prevalent** in the U.S. But some people will not buy them because of health **concerns**. As a result, many companies avoid GM foods although there is no physical evidence that they are unhealthy.

Time Will Tell

Genetically modified foods have not hurt anyone. Most genetic researchers think that if troubles do crop up, they will be manageable. It is important to keep researching GM foods because these types of foods can better fight the world's chronic hunger problems.

Make Connections

Talk about the advantages and disadvantages of GM foods. **ESSENTIAL QUESTION**

Would you buy GM foods? **TEXT TO SELF**

Mapping Specialists

Reread

When you read informational text you may come across facts and details that are new to you. As you read "Food Fight," reread the difficult sections to make sure you understand and remember new information in the text.

Find Text Evidence

You may not be sure you understand why genetically modified foods are created. Reread the last two paragraphs on page 223 in "Food Fight," which explain the goal behind genetically modified foods.

page 223

An incredible thing is happening to our food. Some scientists are using a technique called genetic modification to make superior food crops. It involves altering a seed's genes. Genes are the "instruction codes" that all living things have inside their cells. A seed's genetic code sets what **characteristics** it will **inherit** when it grows into a plant. These could mean how big it will grow and the nutrients it will contain.

For thousands of years, farmers made crops better by crossbreeding plants. They would add pollen from the sweetest melon plants to the flowers of plants that produced the biggest melons. This would make new plants with big, sweet melons. But this process does not always work. The cycle of crossbreeding can take years to get good results.

But advances in gene science have created amazing shortcuts. Using new tools, scientists can put a gene from one living thing into another.

That living thing could be a plant, a bacterium, a virus, or even an animal. These foods are called genetically modified foods, or GM foods. The goal of GM foods is to create foods that can survive insects or harsh conditions or can grow faster. But are these **advancements** in **agriculture** good for us?

I read that the goal is to create foods that can survive insects and harsh conditions. From this, I can draw the inference that scientists are trying to help farmers.

Your Turn

Why do some people think that GM foods are not a good idea? Reread "Safety Issues" on page 225 to answer the question. As you read, remember to use the strategy Reread.

Author's Point of View

Authors have a position or point of view about the topics they write about. Look for details in the text, such as the reasons and evidence the author chooses to present. This will help you to figure out the author's point of view.

 Find Text Evidence

When I reread page 224 of "Food Fight," I can identify details in the text that explain and support the author's position or attitude. Then I can figure out the author's point of view.

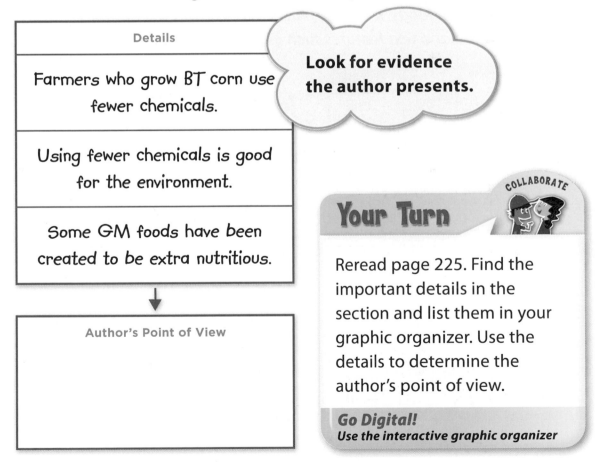

Details
Farmers who grow BT corn use fewer chemicals.
Using fewer chemicals is good for the environment.
Some GM foods have been created to be extra nutritious.

Look for evidence the author presents.

↓

Author's Point of View

Your Turn COLLABORATE

Reread page 225. Find the important details in the section and list them in your graphic organizer. Use the details to determine the author's point of view.

Go Digital!
Use the interactive graphic organizer

Persuasive Article

"Food Fight" is a persuasive article.

A persuasive article:

- Is nonfiction stating the author's opinion on a topic.
- Provides facts and examples to persuade the reader to agree with the author's opinion.
- May include text features such as charts and maps.

 Find Text Evidence

"Food Fight" is a persuasive article. It states the author's opinion about GM foods. It includes facts that support the author's opinion and text features such as headings, charts, and maps.

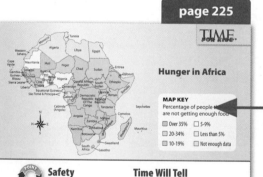

page 225

TIME

Hunger in Africa

MAP KEY
Percentage of people who are not getting enough food
- ☐ Over 35%
- ☐ 5-9%
- ☐ 20-34%
- ☐ Less than 5%
- ☐ 10-19%
- ☐ Not enough data

Text Feature

Maps Maps show geographic locations of specific areas of the world. They usually include a map key and a compass rose.

 Safety Issues

Many people have **disagreed** with the idea that GM foods are a good idea. They worry GM foods will hurt the environment and humans. One concern is that plants with new genes will crossbreed with weeds to make pesticide-resistant weeds. Another concern is that GM foods may trigger allergies.

Genetically modified crops are **prevalent** in the U.S. But some people will not buy them because of health concerns. As a result, many companies avoid GM foods although there is no physical evidence that they are unhealthy.

Time Will Tell

Genetically modified foods have not hurt anyone. Most genetic researchers think that if troubles do crop up, they will be manageable. It is important to keep researching GM foods because these types of foods can better fight the world's chronic hunger problems.

Make Connections
Talk about the advantages and disadvantages of GM foods. **ESSENTIAL QUESTION**

Would you buy GM foods? **TEXT TO SELF**

225

Your Turn COLLABORATE

Find two text features in "Food Fight." Tell what information you learned from each feature.

Greek Roots

Knowing Greek roots can help you figure out the meanings of unfamiliar words. Here are some common Greek roots that may help you as you read "Food Fight."

gen = race, kind
phys = nature
chron = time

Find Text Evidence

When I read the word cycle *on page 223, I know the Greek root* cycl *means* circle. *The word* process *also provides a clue.* Cycle *means* a series of events that recur regularly.

But this process does not always work. The cycle of crossbreeding can take years to get good results.

Your Turn

COLLABORATE

Use Greek roots and context clues to find the meanings of the following words from "Food Fight."

gene, *page 223*
physical, *page 225*
chronic, *page 225*

John Lamb/Stone/Getty Images

Readers to...

Writers have a purpose, or reason, for what they write. They write to entertain, inform, or persuade. Writers also think about their audience, the people who will read their writing. Reread the excerpt from "Food Fight."

Purpose and Audience

Identify the **purpose** and **audience** using clues from the text. Why has the writer written this piece? Who is the audience?

Expert Model

Scientists believe the new techniques can make crops with a resistance to pests and disease. Bt corn is a genetically modified corn. It has an insect-killing gene that comes from a bacterium. Farmers who grow Bt corn can use fewer chemicals while they grow their crops. That is good for the farmer and the environment.

Writers

Sam wrote about science camp. Read Sam's revisions to a section of his essay.

SUMMER OF SCIENCE

Today is my last day at science

camp. It ~~is~~ an ~~awesome~~ way to spend
        was   exciting

my summer. The camp ~~begun~~ with
                          began

nature week. We studied different

kinds of plants and insects in the

woods. I learned that some ~~speceis~~ of (SP)

grasshoppers have ears on ~~there~~ front
                                their

legs. Later in the summer we studied

marine biology. I ~~feeled~~ a dolphin's
                      felt

skin at the aquarium. it was very

slippery and smooth.

Editing Marks

⌒ Switch order.

∧ Add.

⌄ Add a comma.

𝒮 Take out.

(SP) Check spelling.

≡ Make a capital letter.

Grammar Handbook

Irregular Verbs
See page 462.

Your Turn

COLLABORATE

☑ Identify Sam's audience and purpose.
☑ Identify the irregular verbs he corrected.
☑ Tell how other revisions improved his writing.

Go Digital!
Write online in Writer's Space

Fact or Fiction?

The Big Idea

How do different writers treat the same topic?

Star light, star bright,
The first star I see tonight;
I wish I may, I wish I might,
Have the wish I wish tonight.

Essential Question

Why do we need government?

Go Digital!

JUSTICE FOR ALL

The justice system is an example of government at work. Judges are appointed by elected officials or are elected directly by the voters. Juries are made up of citizens who listen to the evidence presented by both sides.

▶ What is another example of government at work?

▶ What services does your state government provide?

▶ What might happen if there were no government?

Talk About It COLLABORATE

Write words that describe the different roles that government plays. Then talk to your partner about why we need government.

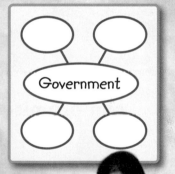

Government

Vocabulary

Use the picture and the sentences to talk with a partner about each word.

amendments

One of the **amendments** to the Constitution gave women the right to vote.

Why do we need amendments?

commitment

The two boys made a **commitment** to practice their song for the talent show.

What is a commitment you have made?

compromise

Sam and his dad agreed to **compromise** on when Sam would mow the lawn.

Describe a situation in which you felt you had to compromise.

democracy

In a **democracy**, it is important for people to vote during an election.

How is a democracy different from a government ruled by a King?

eventually

Grace knew that **eventually** the rain would finally stop.

What is a place that you would like to visit eventually?

legislation

Congress passed **legislation** protecting workers who are injured on the job.

Why might it be important to pass new legislation?

privilege

My grandmother feels that going out to dinner with her family once a week is a **privilege** she deserves.

What is a privilege you wish you had?

version

For this **version** of the movie *Cinderella*, we had to wear 3-D glasses.

What are some things that may have different versions?

Your Turn

COLLABORATE

Pick three words. Write three questions for your partner to answer.

Go Digital! Use the online visual glossary

A WORLD WITHOUT RULES

Essential Question

Why do we need government?

Read how government and laws help to protect us every day.

You may sometimes wonder if rules were made to keep you from having fun and to tell you what to do. But what if we had no rules at all? Nobody would tell you what to do ever again! Sounds great, right? Well, let's see what it's like to inhabit a world without rules. You just might change your mind!

A Strange Morning

Let's start at home. Your alarm clock goes off. Why hurry? Without rules you don't have to go to school. **Eventually** you wander downstairs and find your little brother eating cookies in the kitchen. Since there are no rules, you can have cookies for breakfast! But you wonder if you should have something sensible like a bowl of cereal. You reach a **compromise** (KOM•pruh•mighz) and crumble the cookies over your cereal. In this new world, you will not have to brush your teeth anymore. Of course, the next time you see the dentist, you may have a cavity.

A Community in Confusion

Now, you step outside. You decide to go to the playground because there's no law saying you have to go to school. No crossing guard stands at the corner to help you across the street. Without traffic laws, cars zip by at an alarming speed honking at each other, and there is not a police officer in sight. There is no safe alternate way to cross the street. Besides, once you see the playground, you may decide it is not worth the risk of getting hit by a car. Broken swings dangle from rusty chains. Trash cans overflow with plastic bottles, snack wrappers, and paper bags. A huge tree branch lies across the sliding board. As a result of all state and federal services being gone, nobody is in charge of taking care of the playground.

239

Now think about trying to do all the other things you love. Want to go to the beach? The lifeguards will not be there to keep you safe. Want to play soccer in the park? Your state and local governments are not around to maintain the parks, so you'll never find a place to play. Feel like eating lunch outside? As a result of pollution, the air quality is so bad that you will probably have to wear a gas mask every day.

Have you ever thought about our country being invaded by another country? Remember, the government runs the army. Without the government, there is no army to protect us if another country decided to take over our country.

Back to Reality

Thankfully, that **version** of our world isn't real. We live in a **democracy** (di•MOK•ruh•see) where we have the **privilege** (PRIV•uh•lij) of voting for the people that we want to run the country. Our elected government passes **legislation** (lej•is•LAY•shuhn), or laws, meant to help and protect us. If the country outgrows an old law, then the government can pass **amendments** to the law. Community workers such as crossing guards, police officers, and lifeguards all work to keep you safe, while government agencies such as the Environmental Protection Agency have made a **commitment** to inspect the air and water for pollution. And don't forget the armed forces, which were created to protect our nation.

Our government and laws were designed to keep you safe and ensure you are treated as fairly as everyone else. Without them, the world would be a different place.

240

Make Connections

Talk about how government helps us maintain order and helps preserve our freedom. **ESSENTIAL QUESTION**

What are some ways that the government protects you every day? **TEXT TO SELF**

Ask and Answer Questions

When you read informational text, you may come across facts and ideas that are new to you. Stop and ask yourself questions to help you understand and remember the information. Then read the text closely to find the answers.

Find Text Evidence

When you first read the "Back to Reality" section in a "A World Without Rules," you may have asked yourself what role the Environmental Protection Agency has in keeping people safe.

page 240

Back to Reality

 Thankfully, that **version** of our world isn't real. We live in a **democracy** (di•MOK•ruh•see) where we have the **privilege** (PRIV•uh•lij) of voting for the people that we want to run the country. Our elected government passes **legislation** (lej•is•LAY•shuhn), or laws, meant to help and protect us. If the country outgrows an old law, then the government can pass **amendments** to the law. Community workers such as crossing guards, police officers, and lifeguards all work to keep you safe, while government agencies such as the Environmental Protection Agency have made a **commitment** to inspect the air and water for pollution. And don't forget the armed forces, which were created to protect our nation.

 Our government and laws were designed to keep you safe and ensure you are treated as fairly as everyone else. Without them, the world would be a different place.

As I read on, I found the answer to my question. The Environmental Protection Agency's role is to inspect our air and water and make sure that they are clean.

Your Turn

COLLABORATE

Think of two questions about "A World Without Rules." Then read to find the answers. As you read, remember to use the strategy Ask and Answer Questions.

Cause and Effect

Authors use text structure to organize the information in a nonfiction work. Cause and effect is one kind of text structure. A cause is why something happens. An effect is what happens. Signal words such as *because, so, since,* and *as a result* can help you identify cause-and-effect relationships.

 Find Text Evidence

When I reread the section "A Strange Morning" on page 239, I will look for causes and effects. I will also look for signal words.

Cause	→	Effect
Without rules	→	You don't have to go to school
Without rules	→	You can have cookies for breakfast.
You don't have to brush your teeth.	→	You may get a cavity.

The effect is what happens as a result of an action.

Your Turn COLLABORATE

Reread "A Community in Confusion" on pages 239–240. Identify the causes and effects. List them in the graphic organizer.

Go Digital!
Use the interactive graphic organizer

Narrative Nonfiction

"A World Without Rules" is narrative nonfiction.

Narrative nonfiction:

- Is told in the form of a story.
- May express the author's opinion about the subject.
- Presents facts and includes text features.

Find Text Evidence

"A World Without Rules" is narrative nonfiction. The author tells a story and includes text features. The author also expresses an opinion and supports it with facts and examples.

page 239

You may sometimes wonder if rules were made to keep you from having fun and to tell you what you to do. But what if we had no rules at all? Nobody would tell you what to do ever again! Sounds great, right? Well, let's see what it's like to inhabit a world without rules. You just might change your mind!

A Strange Morning

Let's start at home. Your alarm clock goes off. Why hurry? Without rules you don't have to go to school. **Eventually** you wander downstairs and find your little brother eating cookies in the kitchen. Since there are no rules, you can have cookies for breakfast. But you wonder if you should have something sensible like a bowl of cereal. You reach a **compromise** (KOM•pruh•mighz) and crumble the cookies over your cereal. In this new world, you will not have to brush your teeth anymore. Of course, the next time you see the dentist, you may have a cavity.

A Community in Confusion

Now, you step outside. You decide to go to the playground because there's no law saying you have to go to school. No crossing guard stands at the corner to help you across the street. Without traffic laws, cars zip by at an alarming speed honking at each other, and there is not a police officer in sight. There is no safe alternate way to cross the street. Besides, once you see the playground, you may decide it is not worth the risk of getting hit by a car. Broken swings dangle from rusty chains. Trash cans overflow with plastic bottles, snack wrappers, and paper bags. A huge tree branch lies across the sliding board. As a result of all state and federal services being gone, nobody is in charge of taking care of the playground.

239

Text Features

Boldface Words Boldface words show key words in the text.

Pronunciations Pronunciations show how to sound out unfamiliar words.

Your Turn

COLLABORATE

Reread "A World Without Rules." What is the author's opinion of government? Find text evidence to support your answer.

Latin Roots

Knowing Latin roots can help you figure out the meanings of unfamiliar words. Look for these Latin roots as you read "A World Without Rules."

dent = tooth   *commun* = common   *spect* = look

 ### Find Text Evidence

When I read the third paragraph on page 239, in the section "A Community in Confusion," I see the word alternate. *The Latin root* alter *means other. This will help me figure out what* alternate *means.*

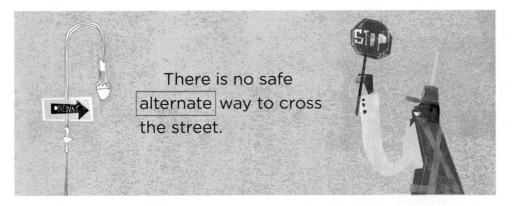

There is no safe alternate way to cross the street.

Your Turn

COLLABORATE

Use context clues and Latin roots to figure out the meanings of these words in "A World Without Rules."

dentist, *page 239*

community, *page 240*

inspect, *page 240*

RG Roth

Readers to...

Writers build strong paragraphs by stating the main idea in a topic sentence. They include supporting sentences that give more information about the main idea. Reread the excerpt from "A World Without Rules" below.

Expert Model

Strong Paragraphs

Identify the **topic sentence** and the **supporting sentences.** How do the facts and examples support the main idea?

We live in a democracy where we have the privilege of voting for the people that we want to run the country. Our elected government passes legislation, or laws, meant to help and protect us. If the country outgrows an old law, then the government can pass amendments to the law. Community workers such as crossing guards, police officers, and lifeguards all work to keep you safe, while government agencies such as the Environmental Protection Agency have made a commitment to inspect the air and water for pollution.

STOP

ONE WAY

Writers

Stefan wrote an essay about rules. Read Stefan's revisions to one section of his essay.

Editing Marks

⎍ Switch order.

∧ Add.

∧ Add a comma.

✐ Take out.

(SP) Check spelling.

≡ Make a capital letter.

Grammar Handbook

Pronouns and Antecedents See page 463.

Student Model

Rules

Rules ~~are good~~. Rules help keep
may not be fun, but they are helpful.
∧
Can you imagine how crazy things
would get if people just did what they wanted?
our society orderly. Rules also keep
∧

you safe. For example, pool rules

make sure kids do not slip on the

deck or hurt themselves while diving.

~~I think there are too many rules at~~

~~pools sometimes~~. Finally, rules help
                                  you
you know how ~~they~~ should behave. If
                ∧
        them
you follow ~~rules~~, you'll know that you
              ∧

are acting the right way for a certain

place or situation.

Your Turn

- ✔ Identify Stefan's topic sentence.
- ✔ Identify a pronoun and its antecedent.
- ✔ Tell how Stefan's revisions made his paragraph stronger.

Go Digital!
Write online in Writer's Workspace

Essential Question
Why do people run for public office?

Go Digital!

TAKING A STAND

During the Great Depression, there were a number of hunger marches. All over the country, people marched to protest unemployment, lack of health insurance, and the state of the economy. This photograph was taken in 1932 during one of the marches.

NATIONAL HUNGER MARCH 1932

▶ If you had been a politician running for office in 1932, what would you have told the voters?

▶ What qualities does a good leader need?

Talk About It COLLABORATE

Write words that you have learned about leadership. Then talk with a partner about what would make you want to run for public office.

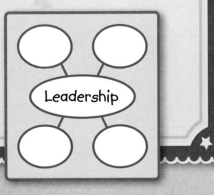

Leadership

Vocabulary

Use the picture and the sentences to talk with a partner about each word.

accompanies

Jake's dog **accompanies** him on car rides.

What is a synonym for accompanies?

campaign

The woman signed up to work on Mr. Baker's **campaign** for the state senate.

What are some activities a politician does during a campaign?

governor

The **governor** spoke at a town hall meeting about the state budget.

What are some ways a governor can help the people in his or her state?

intend

Does the mouse **intend** to eat the grape?

What do you intend to do tomorrow?

opponent

Laili and her brother beat the online **opponent** they were playing against in the video game.

What is an antonym for opponent?

overwhelming

The number of books that Todd had to carry was **overwhelming**.

What is a synonym for overwhelming?

tolerate

Polar bears can **tolerate** extremely cold water.

What kind of animal can tolerate living in a hot desert climate?

weary

The firefighter was **weary** and needed to rest after fighting a fire for 10 hours.

Why else might someone be weary?

COLLABORATE

Your Turn

Pick three words. Write three questions for your partner to answer.

Go Digital! *Use the online visual glossary*

The TimeSpecs 3000

Essential Question

Why do people run for public office?

Read how Miguel makes a decision to run for class president.

September 15

Dear Grandpa,

I just got back from our class field trip to Washington, D.C., and I have a lot to tell you. Going to Washington helped me decide to run for class president.

I owe it all to your invention, the TimeSpecs 3000! In a nutshell, it helped me get some helpful advice about my problem. I **intend** to tell you everything when I visit Saturday, but for now I've pasted my field notes into this e-mail, so you can understand how well your invention worked.

FIELD NOTES: **DAY 1**

I use the TimeSpecs 3000 at the Washington Monument. Our guide **accompanies** us everywhere, and while she's talking I put on the specs. The design needs tweaking because my friend Ken whispered, "Nerdy shades, dude!"

Immediately, I'm seeing the monument in the past. I am watching the ceremony when they laid the cornerstone in 1848, and everybody's wearing large hats and funny, old-fashioned clothes. When I take off the TimeSpecs 3000, I realize my class is heading to lunch so I run after them.

253

We're back on the National Mall, which is nothing like Brookfield's mall with all its stores. This mall is outside and has a long reflecting pool. My teacher is finding it hard to **tolerate** some of my classmates' immature behavior, which includes running around throwing pebbles in the reflecting pool. I'm getting kind of **weary** of all the noise, and I'd rather learn about history on my own. So I put on the TimeSpecs 3000 and check out the Lincoln Memorial.

I see how dignified Lincoln's statue looks and wonder if I could ever help people like he did. This starts me thinking again about whether I should run for class president. Suddenly, right out of the blue, I hear this voice. "Excuse me, young man. You're thinking of running for president?" I look up and realize that Lincoln's statue is talking to me. It's so **overwhelming** that I stand there speechless for a minute.

Finally, I stammer, "President . . . Lincoln?"

"Maybe you should first run for mayor of your town," the statue says. "Or perhaps for **governor**? Once you get the hang of being in public office, you could run for president."

"Actually, it's for president of my 4th grade class," I say.

The giant statue nods. "That's an excellent start."

I figure while I have Lincoln's ear, I should get some advice. "I have a problem. I hate writing and giving speeches, and my **opponent**, Tommy, is great at both things."

"What kind of **campaign** would you run?" Lincoln asks.

"I have lots of ideas for our school," I tell him. "For instance, I want our school to use fruits and vegetables from the local farmers' market in the cafeteria. I also want to start a book drive for our school library."

"There's your speech," he says. "Tell people your ideas with honesty, integrity, and enthusiasm, and you can't possibly go wrong."

"Thanks, Mr. President," I say. "I think I can do that!"

Grandpa, I can't wait to see you on Saturday because I have to tell you about our visit to the Natural History Museum.

Your grandson and future class president,
Miguel

P.S. I would advise not wearing the TimeSpecs 3000 while looking at dinosaur bones.

Make Connections

Talk about why Miguel decides to run for class president. **ESSENTIAL QUESTION**

What would you do for your school if you were class president? **TEXT TO SELF**

Chris Boyd

Make Predictions

When you read you can use details from the story to make predictions about what you think will happen. As you read "The TimeSpecs 3000," make predictions about the story and confirm or revise them.

🔍 Find Text Evidence

What kind of invention did you predict the TimeSpecs 3000 was? Go back and reread the beginning of the e-mail on page 253. What details helped you to make your prediction?

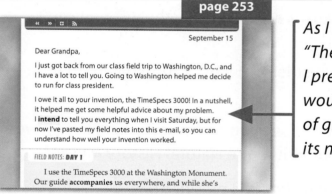

page 253

September 15

Dear Grandpa,

I just got back from our class field trip to Washington, D.C., and I have a lot to tell you. Going to Washington helped me decide to run for class president.

I owe it all to your invention, the TimeSpecs 3000! In a nutshell, it helped me get some helpful advice about my problem. I **intend** to tell you everything when I visit Saturday, but for now I've pasted my field notes into this e-mail, so you can understand how well your invention worked.

FIELD NOTES: **DAY 1**

I use the TimeSpecs 3000 at the Washington Monument. Our guide **accompanies** us everywhere, and while she's

As I read page 253 of "The TimeSpecs 3000," I predicted the invention would be a special kind of glasses because of its name.

Your Turn

COLLABORATE

Read page 255 of "The TimeSpecs 3000." What clues did you find in the text that led you to predict Abraham Lincoln's words would help Miguel solve his problem?

Point of View

The narrator's point of view is how the narrator thinks or feels about characters or events in the story. A story can have a first-person narrator or a third-person narrator.

 Find Text Evidence

When I read page 253 of "The TimeSpecs 3000," I learn that a boy is writing an e-mail to his grandfather. I see the pronouns I, me *and* my *so I know this story has a first-person narrator. I can find details in the story to find the narrator's point of view.*

Details
The narrator was weary of the noise his classmates made and wanted to learn history on his own.
The narrator wonders if he could ever help people like Lincoln did.

↓

Point of View
The narrator is a fourth-grader excited by history. He is unsure if he should run for class president.

Your Turn COLLABORATE

Find other details from "The TimeSpecs 3000" that tell you the narrator's point of view. Put the information in the graphic organizer.

Go Digital!
Use the interactive graphic organizer

Fantasy

"The TimeSpecs 3000" is a fantasy.

A fantasy:

- Is a type of fiction story.
- Has characters, settings, or events that could not exist in real life.
- Usually includes illustrations.

Find Text Evidence

"The TimeSpecs 3000" is a fantasy. The character of Miguel is realistic, but when he uses the TimeSpecs 3000 he is able to see things that happened in the past. Also, some of the illustrations depict events that could not happen in real life.

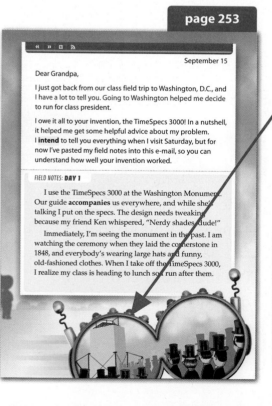

page 253

September 15

Dear Grandpa,

I just got back from our class field trip to Washington, D.C., and I have a lot to tell you. Going to Washington helped me decide to run for class president.

I owe it all to your invention, the TimeSpecs 3000! In a nutshell, it helped me get some helpful advice about my problem. I **intend** to tell you everything when I visit Saturday, but for now I've pasted my field notes into this e-mail, so you can understand how well your invention worked.

FIELD NOTES: **DAY 1**

I use the TimeSpecs 3000 at the Washington Monument. Our guide **accompanies** us everywhere, and while she's talking I put on the specs. The design needs tweaking because my friend Ken whispered, "Nerdy shades, dude!"

Immediately, I'm seeing the monument in the past. I am watching the ceremony when they laid the cornerstone in 1848, and everybody's wearing large hats and funny, old-fashioned clothes. When I take off the TimeSpecs 3000, I realize my class is heading to lunch so I run after them.

Illustrations Illustrations show the events of the story. Here we can see what the world looks like viewed through the TimeSpecs 3000.

COLLABORATE

Your Turn

Find two more examples in the text that show "The TimeSpecs 3000" is a fantasy. Discuss these examples with your partner.

Idioms

An idiom is a phrase or expression whose meaning cannot be understood from the separate words in it. If you are not sure of the meaning of an idiom, look at the surrounding phrases and sentences to help you figure it out.

Find Text Evidence

The phrase in a nutshell *on page 253 is an idiom. I know it does not mean that something is really inside the shell of a nut. Miguel says that his grandfather's invention helped him reach a decision. I think that* in a nutshell *means to summarize or say briefly.*

I owe it all to your invention, the TimeSpecs 3000! In a nutshell, it helped me get some helpful advice about my problem.

Your Turn

COLLABORATE

Use context clues to figure out the meanings of the following idioms in "The TimeSpecs 3000."

out of the blue, *page 254*

get the hang of, *page 254*

have Lincoln's ear, *page 255*

Chris Boyd

Readers to...

Writers use dialogue to show what a character is feeling and thinking. Writers also pay attention to how the character says the dialogue. Reread the excerpt from "The TimeSpecs 3000" below.

Expert Model

Develop Character

Identify the **dialogue.** How do these words help you understand how Miguel feels as he talks to Abraham Lincoln?

"I have a problem. I hate writing and giving speeches, and my opponent, Tommy, is great at both things."

"What kind of campaign would you run?" Lincoln asks.

"I have lots of ideas for our school," I tell him. "For instance, I want our school to use fruits and vegetables from the local farmers' market in the cafeteria. I also want to start a book drive for our school library."

"There's your speech," he says. "Tell people your ideas with honesty, integrity, and enthusiasm, and you can't possibly go wrong."

Writers

Nina wrote a fantasy. Read Nina's revisions to a section of her story.

Editing Marks

⌐⌐ Switch order.

∧ Add.

⋏ Add a comma.

✍ Take out.

(SP) Check spelling.

≡ Make a capital letter.

Grammar Handbook

Types of Pronouns
See page 463.

Student Model

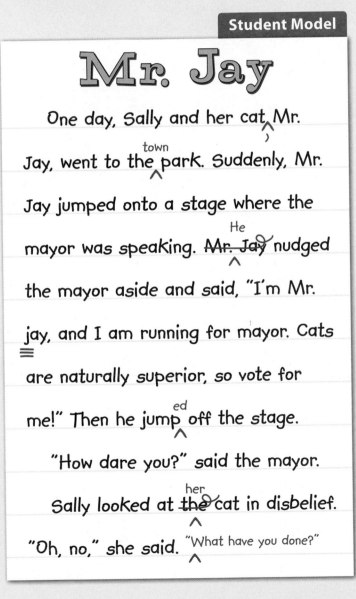

Mr. Jay

One day, Sally and her cat, Mr.

                    town
Jay, went to the ∧ park. Suddenly, Mr.

Jay jumped onto a stage where the
                                He
mayor was speaking. ~~Mr. Jay~~ nudged
                      ∧
the mayor aside and said, "I'm Mr.

jay, and I am running for mayor. Cats
≡
are naturally superior, so vote for
                   ed
me!" Then he jump∧ off the stage.

"How dare you?" said the mayor.
                     her
Sally looked at ~~the~~ cat in disbelief.
                  ∧
"Oh, no," she said. "What have you done?"
                   ∧

Your Turn

COLLABORATE

☑ Identify **dialogue** in the story. How does it help Nina develop characters?

☑ Identify the types of pronouns that she used in her writing.

☑ Tell how the revisions improved Nina's writing.

Go Digital!
Write online in Writer's Workspace

261

Essential Question
How do inventions and technology affect your life?

Go Digital!

Changing LIVES

For some people, new inventions and advanced technology provide a way to fulfill their dreams. The man in the photo is able to compete in the Paralympics now that he has an artificial leg that allows him to run long distances.

▶ How do you think inventions and technology help to make our lives better?

▶ What technology or invention do you rely on the most? Why?

Talk About It

Write words that describe how inventions and technology affect your life. Then talk with a partner about an invention that you would like to design that would have a big impact on your life.

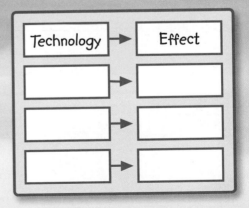

Technology →	Effect

Vocabulary

Use the picture and the sentences to talk with a partner about each word.

decade

The company celebrated a **decade** of business, honoring ten years of work.

What year will it be in a decade?

directing

The police officers are **directing** traffic.

If you were a crossing guard, what would you be directing students to do?

engineering

I think the beautiful Golden Gate Bridge is an amazing feat of **engineering**.

What is another structure that was built using the science of engineering?

gleaming

The shiny bar of gold lay **gleaming** on the red velvet.

What is an antonym for gleaming?

scouted

The boy used binoculars as he **scouted** the best place to find whales.

What is a synonym for scouted?

squirmed

The pig wiggled and **squirmed** in the girl's arms.

If someone squirmed while watching a play, how did that person probably feel?

technology

In the early 1900s, the telephone was considered new **technology**.

What are some examples of new technology today?

tinkering

Mr. Lan likes **tinkering** with and fixing old clocks.

What do you like tinkering with?

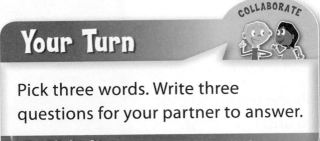

COLLABORATE

Your Turn

Pick three words. Write three questions for your partner to answer.

Go Digital! *Use the online visual glossary*

A Telephone Mix-Up

Essential Question

How do inventions and technology affect your life?

Read how a telephone brings change to the lives of Meg and her father.

"By tomorrow afternoon there will be eight telephones right here in Centerburg, Ohio, and one of them will be ours!" Dr. Ericksen said to his daughter, Meg. "I predict that before this **decade** is over, in just another five years, there could be a hundred! That's how fast I foresee this **technology** will spread! When people need help, they'll call me on the telephone. Envision how many lives it will save! Picture all the amazing benefits!"

Meg realized that not everyone thought the telephone was an **engineering** marvel. She had heard people say that telephones were a useless invention. A few others felt the newfangled machine would open up a Pandora's box of troubles, causing people to stop visiting each other and writing letters.

Despite the concerns of some people, progress marched on. Just weeks earlier, Centerburg's first telephone had been installed in Mr. Kane's general store, another was put in at the hotel, and yet another at the newspaper office. Mrs. Kane was the town's first switchboard operator, **directing** incoming calls to the correct lines.

The next morning, Meg wrote "October 9, 1905" on the top of her slate with chalk while she **squirmed** in her seat, wishing that the long school day was over.

Tristan Elwell

Walking home that afternoon, Meg **scouted** the street, looking for the tall wooden poles that were going up weekly. Thick wire linked one pole to another, and Meg imagined how each wire would carry the words of friends and neighbors, their conversations zipping over the lines bringing news, birthday wishes, and party invitations.

As Meg hurried into the house, she let the screen door slam shut behind her. There on the wall was the **gleaming** wooden telephone box with its heavy black receiver on a hook. Her father was smiling broadly while **tinkering** with the shiny brass bells on top. "Isn't it a beauty?" he asked. "Have you ever seen such magnificence?"

Suddenly the telephone jangled loudly, causing both Ericksens to jump.

Meg laughed as her father picked up the receiver and shouted, "Yes, hello, this is the doctor!"

"Again please, Mrs. Kane! There's too much static" Dr. Ericksen shouted. "I didn't get the first part. Bad cough? Turner farm?"

"Can I go, Father?" Meg asked as Dr. Ericksen returned the receiver to the hook.

"Absolutely," he said, grabbing his medical kit and heading outside where his horse and buggy waited.

When they got to the farm, they found Mr. Turner walking toward the barn.

"Jake, I got here as quick as I could," Dr. Ericksen said. "Is it Mrs. Turner? Little Emma?"

"You?" Jake Turner looked confused, but he gestured them toward the barn.

There they found a baby goat curled near its mother. The baby snorted, coughed, and looked miserable.

"Jake, I'm no vet!" said Dr. Ericksen. "You need Dr. Kerrigan."

"I was wondering why you showed up instead. I reckon there was a mix-up."

"Apparently so," Dr. Ericksen laughed. "When I get back I'll send Dr. Kerrigan."

As years passed the telephone proved to be very useful to the town of Centerburg, but there was always the occasional mix-up. It became common among the Ericksens to refer to a missed communication as "another sick goat."

Make Connections

How did the invention of the telephone affect the town of Centerburg? **ESSENTIAL QUESTION**

Think of an invention and tell how it has affected your life. **TEXT TO SELF**

Make Predictions

When you read, use text clues from the story to help you make predictions about what will happen next. As you continue to read, you can confirm or revise your predictions.

Find Text Evidence

How did you predict the people of Centerburg would react to the telephone? What helped you to confirm your prediction? Reread page 267 of "A Telephone Mix-Up."

page 267

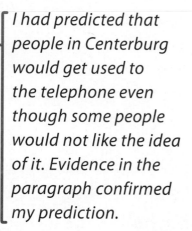

would open up a Pandora's box of troubles, causing people to stop visiting each other and writing letters.

Despite the concerns of some people, progress marched on. Just weeks earlier, Centerburg's first telephone had been installed in Mr. Kane's general store, another was put in at the hotel, and yet another at the newspaper office. Mrs. Kane was the town's first switchboard operator, **directing** incoming calls to the correct lines.

The next morning, Meg wrote "October 9, 1905" on the top of her slate with chalk while she **squirmed** in her seat, wishing that the long school day was over.

I had predicted that people in Centerburg would get used to the telephone even though some people would not like the idea of it. Evidence in the paragraph confirmed my prediction.

COLLABORATE

Your Turn

What text clues did you find that helped you predict that the phone static would cause a mix-up? As you read remember to use the strategy, Make Predictions.

Point of View

The narrator's point of view tells how the narrator thinks or feels about characters or events in the story. A story can have a first-person narrator or a third-person narrator.

 Find Text Evidence

When I read page 267 of "A Telephone Mix Up," I see that the narrator uses the pronouns *he* and *she* when the narrator tells what Meg and her father are thinking. This story has a third-person narrator. I can find details in the story about the narrator's point of view.

Details
The narrator tells us what Meg's father says about the telephone. "Picture all the amazing benefits!"
The narrator states: "Despite the concerns of some people, progress marched on."

↓

Point of View
The narrator thinks the telephone will be a useful invention.

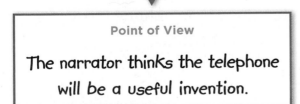

Your Turn

Reread "A Telephone Mix-Up." Find other details from the story that tell you the narrator's point of view. Use the graphic organizer to list the details.

Go Digital!
Use the interactive graphic organizer

Historical Fiction

"A Telephone Mix-Up" is historical fiction.

Historical fiction:

- Has realistic characters, events, and settings.
- Is set in the past and based on real events.
- Contains dialogue.

Find Text Evidence

"A Telephone Mix-Up" is historical fiction. A family is getting a telephone at a time in history when telephone service was first made available to many communities. The story has realistic characters, events, and settings, and it includes dialogue.

page 267

"By tomorrow afternoon there will be eight telephones right here in Centerburg, Ohio, and one of them will be ours!" Dr. Ericksen said to his daughter, Meg. "I predict that before this **decade** is over, in just another five years, there could be a hundred! That's how fast I foresee this **technology** will spread! When people need help, they'll call me on the telephone. Envision how many lives it will save! Picture all the amazing benefits!"

Meg realized that not everyone thought the telephone was an **engineering** marvel. She had heard people say that telephones were a useless invention. A few others felt the newfangled machine would open up a Pandora's box of troubles, causing people to stop visiting each other and writing letters.

Despite the concerns of some people, progress marched on. Just weeks earlier, Centerburg's first telephone had been installed in Mr. Kane's general store, another was put in at the hotel, and yet another at the newspaper office. Mrs. Kane was the town's first switchboard operator, **directing** incoming calls to the correct lines.

The next morning, Meg wrote "October 9, 1905" on the top of her slate with chalk while she **squirmed** in her seat, wishing that the long school day was over.

Dialogue Dialogue is the conversation that takes place between the characters. Quotation marks enclose dialogue.

Your Turn

COLLABORATE

Find three examples in the text that show "A Telephone Mix-Up" is historical fiction.

Synonyms

As you read "A Telephone Mix-Up," you may come across a word you don't know. Sometimes the author will use another word or phrase that has the same or a similar meaning to the unfamiliar word. Words that have the same or similar meanings are **synonyms**.

Find Text Evidence

As I read the first paragraph of "The Telephone Mix-Up" on page 267, I wasn't sure what the word envision *meant. Then the word* picture *in the next sentence helped me figure out the meaning.*

Envision how many lives it will save! Picture all the amazing benefits!

Your Turn

COLLABORATE

Use synonyms and other context clues to find the meanings of the following words in "A Telephone Mix-Up." Write a synonym and example sentence for each word.

foresee, *page 267*
installed, *page 267*
magnificence, *page 268*

Readers to...

Writers develop the plot of a story by including specific details about the story's setting. Reread the excerpt from "A Telephone Mix-Up" below.

Expert Model

Develop Plot

Identify details that give you clues about the story's **setting**. How do the setting details help you understand the story's **plot**?

Despite the concerns of some people, progress marched on. Just weeks earlier, Centerburg's first telephone had been installed in Mr. Kane's general store, another was put in at the hotel, and yet another at the newspaper office. Mrs. Kane was the town's first switchboard operator, directing incoming calls to the correct lines.

The next morning, Meg wrote "October 9, 1905" on the top of her slate with chalk while she squirmed in her seat, wishing that the long school day was over.

Walking home that afternoon, Meg scouted the street, looking for the tall wooden poles that were going up weekly.

Writers

Leo wrote about a boy who sees a television for the first time. Read Leo's revisions to one section of his story.

Student Model

Mike's First Television

Mike sat on the floor in front of the big ^wooden^ box with its small glass screen. <u>t</u>here were several knobs and buttons and a little dial.

"Here it is New Year's Day ^1955^ and ^we^ I are going to see pictures on a television!" Mike's dad said. "This will be a day you will tells your grandchildren about!" Then he turns^ed^ a knob. Slowly a light flickered on the screen.

Editing Marks

⌐⌐ Switch order.

∧ Add.

∧ Add a comma.

⅃ Take out.

(sp) Check spelling.

≡ Make a capital letter.

Grammar Handbook

Pronoun-Verb Agreement See page 464.

Your Turn

COLLABORATE

☑ How do the setting details Leo used help develop the plot?

☑ Identify examples of pronoun-verb agreement that he used in his writing.

☑ Tell how the revisions improved Leo's writing.

Go Digital!
Write online in Writer's Workspace

275

Essential Question

How do you explain what you see in the sky?

Go Digital!

276

The LIGHT SHOW

For centuries, people have come up with stories to explain what they see in the night sky. This photo shows the northern lights, which you will soon read more about. The Inuit who lived on the lower Yukon River believed the northern lights were actually the spirits of animals dancing in the sky.

► If you had lived five hundred years ago, how would you have explained the northern lights?

► What have you observed in the sky at night?

Talk About It

COLLABORATE

Write words that name or describe things that appear in the night sky. Then talk to a partner about stars, planets, comets, and eclipses.

Night Sky

Vocabulary

Use the picture and the sentences to talk with a partner about each word.

astronomer

The **astronomer** pointed out the crater on the planet.

What might an astronomer observe in the night sky?

crescent

The moon tonight looks like a **crescent** and is shaped like a "C."

What other things are shaped like a crescent?

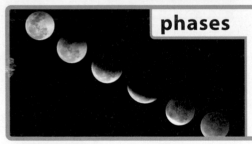

phases

During one of the moon's **phases**, the moon appears to be perfectly round.

Name two phases of the moon.

rotates

The hamster turns and **rotates** his exercise wheel.

What is something else that rotates?

series

This **series** of photographs shows what happened after I watered the flower.

Do you have a favorite series of books?

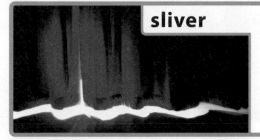

sliver

A thin **sliver** of light showed beneath the curtains.

What is an antonym for sliver?

specific

The boy held up a **specific** kind of orange that is used for making juice.

What is a specific kind of bread that you like best?

telescope

The boy looked through the **telescope** at the boats in the harbor.

What else can you see with a telescope?

Your Turn

COLLABORATE

Pick three words. Write three questions for your partner to answer.

Go Digital! *Use the online visual glossary*

Wonders of the Night Sky

Essential Question

How do you explain what you see in the sky?

Read about what causes some of the sights you see in the sky.

As Earth rotates on its axis, day becomes night. Suddenly, a gallery of lights is revealed! You may see a beautiful crescent moon. Maybe you'll see one of the other phases of the moon. You may even see a series of lights spread across the sky like colored ribbons. For thousands of years, people have loved looking at the night sky. For almost as long, scientists have been trying to explain what they see.

Aurora Borealis

Every few years, an amazing light show is seen in the skies near the North Pole. It is known as "the northern lights," or the **aurora borealis** (uh-RAWR-uh bawr-ee-AL-is). Brilliant bands of green, yellow, red, and blue lights appear in the sky.

People used to believe the lights were caused by sunlight reflecting off polar ice caps. The theory was that when the light bounced back from the caps it created patterns in the sky. In fact, the lights happen because of magnetic attraction.

The sun constantly gives off a stream of electrically charged particles in every direction. These nearly invisible pieces of matter join into a stream called a solar wind. As Earth orbits the sun, solar winds reach Earth's magnetic field. As a result, electric charges occur that are sometimes strong enough to be seen from Earth. These electric charges cause the colorful bands of lights in the sky.

The aurora borealis above
Hammerfest, Norway

Picture Press/Alamy

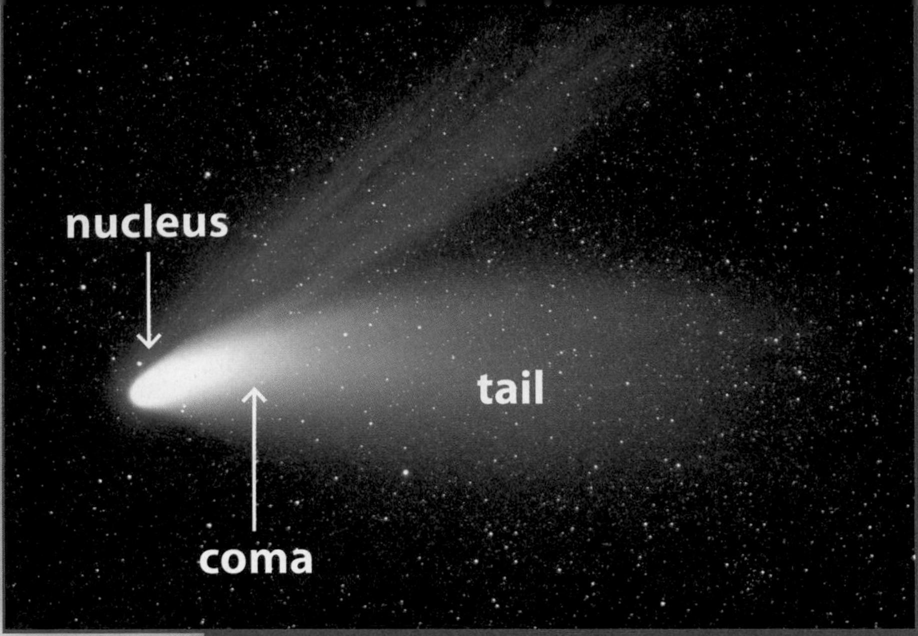

nucleus

tail

coma

This diagram shows the parts of a comet.
Some comets' tails can be millions of miles long.

Comets

Another kind of light you might see move across the night sky is a comet. The word *comet* comes from a Greek word that means, "wearing long hair." It came from the Greek philosopher **Aristotle** (AR-uh-stot-uhl), who thought that comets looked like stars with hair.

Long ago, people feared these mysterious streaks because they believed that they might bring war or sickness to Earth. Today, comets are less scary and mysterious because we know that they are a mixture of rock, dust, ice, and frozen gases that orbit the sun.

Comets move around the sun in an oval-shaped orbit. When a comet comes closer to the sun, the result is that a "tail" of gas and dust is pushed out behind the comet. This long tail is what people see from Earth.

Scientists think comets are some of the oldest objects in space. They can track **specific** comets and predict when they can be seen from Earth again.

Meteors

Have you ever looked up at the sky and seen a shooting star? Those streaks of light are not really stars at all. What we call shooting stars are usually **meteors** (MEE-tee-erz). Meteors are another name for the rocky debris and fragments that enter Earth's atmosphere. Sometimes Earth passes through an area in space with a lot of debris. This is when a meteor shower occurs. You may see hundreds of "shooting stars" on the night of a meteor shower.

The Perseid meteor shower

These days an **astronomer** or anyone with a portable **telescope** can raise new questions about space. What do you see when you look up at the night sky? Whether you look at a **sliver** of the moon or a fantastic light show, you are bound to see something amazing.

Make Connections

Talk about what causes some of the sights in the night sky. **ESSENTIAL QUESTION**

What do you wonder about when you look up at the night sky? **TEXT TO SELF**

Ask and Answer Questions

When you read an informational text, you usually come across new facts and ideas. Asking questions and reading to find the answer can help you understand new information. As you read "Wonders of the Night Sky," ask and answer questions about the text.

Find Text Evidence

When you first read "Wonders of the Night Sky," you may have asked yourself what causes the northern lights.

> **page 281**
>
> **crescent** moon. Maybe you'll see one of the other **phases** of the moon. You may even see a **series** of lights spread across the sky like colored ribbons. For thousands of years, people have loved looking at the night sky. For almost as long, scientists have been trying to explain what they see.
>
> **Aurora Borealis**
>
> Every few years, an amazing light show is seen in the skies near the North Pole. It is known as "the northern lights," or the **aurora borealis** (uh-RAWR-uh bawr-ee-AL-is). Brilliant bands of green, yellow, red, and blue lights appear in the sky.
>
> People used to believe the lights were caused by sunlight reflecting off polar ice caps. The theory was that when the light bounced back from the caps it created patterns in the sky. In fact, the lights happen because of magnetic attraction.
>
> The sun constantly gives off a stream of electrically charged particles in every direction. These nearly invisible pieces of matter join into a stream called a solar wind. As Earth orbits the sun, solar winds reach Earth's magnetic

When I read this section of the text, I found the answer to my question. The northern lights are caused by the sun giving off electrically charged particles.

Your Turn

COLLABORATE

Think of two questions you have about meteors. Reread the section "Meteors" on page 283 and answer your own questions. As you read, remember to use the strategy Ask and Answer Questions.

Cause and Effect

Text structure is the way that authors organize information in a selection. Cause and effect is one kind of text structure. A cause is why something happens. An effect is what happens.

 Find Text Evidence

When I reread the "Aurora Borealis" section on page 281 of "Wonders of the Night Sky," I can look for causes and their effects. Signal words such as cause, because, *and* as a result *tell me that a cause-and-effect relationship is being explained.*

Cause	→	Effect
Sun gives off electrically charged particles.	→	Particles join into a solar wind.
Solar winds reach Earth's magnetic field.	→	As a result, electric charges are seen from Earth.

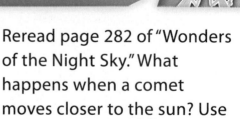

Your Turn

Reread page 282 of "Wonders of the Night Sky." What happens when a comet moves closer to the sun? Use the graphic organizer to list the cause and effect.

Go Digital!
Use the interactive graphic organizer

Expository Text

"Wonders of the Night Sky" is an expository text.

Expository text:
- Explains facts and information about a topic.
- Includes text features.

Find Text Evidence

I know "Wonders of the Night Sky" is an expository text because it gives many facts about the night sky and includes text features. It has boldface words, pronunciations of unfamiliar words, and a diagram.

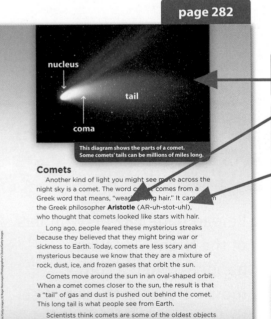

page 282

Comets

Another kind of light you might see move across the night sky is a comet. The word comet comes from a Greek word that means, "wearing long hair." It came from the Greek philosopher **Aristotle** (AR-uh-stot-uhl), who thought that comets looked like stars with hair.

Long ago, people feared these mysterious streaks because they believed that they might bring war or sickness to Earth. Today, comets are less scary and mysterious because we know that they are a mixture of rock, dust, ice, and frozen gases that orbit the sun.

Comets move around the sun in an oval-shaped orbit. When a comet comes closer to the sun, the result is that a "tail" of gas and dust is pushed out behind the comet. This long tail is what people see from Earth.

Scientists think comets are some of the oldest objects in space. They can track **specific** comets and predict when they can be seen from Earth again.

282

Diagram labels: nucleus, tail, coma

This diagram shows the parts of a comet. Some comets' tails can be millions of miles long.

Text Features

Diagrams Diagrams show the parts of something.

Boldface Words Boldface words show key words in the text.

Pronunciations Pronunciations show how to sound out unfamiliar words.

COLLABORATE

Your Turn

Find two text features in "Wonders of the Night Sky." Tell what you learned from each feature.

Context Clues

As you read the information in "Wonders of the Night Sky," you may come across words that you don't know. To figure out the meaning of an unfamiliar word, check the words or phrases near it carefully for clues.

Find Text Evidence

When I read the third paragraph on page 281 of "Wonders of the Night Sky," the phrase light bounced back *helps me figure out what* reflecting *means.*

People used to believe the lights were caused by sunlight reflecting off polar ice caps. The theory was that when the light bounced back from the caps it created patterns in the sky.

Your Turn

Use context clues to find the meaning of the following words in "Wonders of the Night Sky." Write a short definition and example sentence for each word.

particles, *page 281*
mixture, *page 282*
debris, *page 283*

Readers to...

Writers use figurative language such as similes and metaphors to help the reader picture the information being presented. Reread the beginning of page 281 from "Wonders of the Night Sky" below.

Figurative Language

Identify **figurative language** in the text. How does the author use a simile to help the reader visualize the text?

Expert Model

As Earth rotates on its axis, day becomes night. Suddenly, a gallery of lights is revealed! You may see a beautiful crescent moon. Maybe you'll see one of the other phases of the moon. You may even see a series of lights spread across the sky like colored ribbons. For thousands of years, people have loved looking at the night sky. For almost as long, scientists have been trying to explain what they see.

Picture Press/Alamy

288

Writers

Kayla wrote about constellations. Read Kayla's revisions to one section of her text.

Student Model

Editing Marks

⊔ Switch order.

∧ Add.

⌄ Add a comma.

℘ Take out.

(SP) Check spelling.

≡ Make a capital letter.

Grammar Handbook

Possessive Pronouns
See page 465.

Tales of the Night Sky

I've been interest_{ed} in constellations

since I received <del>mine</del> my first telescope

when I was 9. I like to imagene (SP) my

favorite constellation, the big dipper,

like a pitcher full of milk
scooping up the moon. But, to me at

least, some constellations don't look

like what they are supposed to be.

Orion looks more like a row of dots

than
<del>then</del> a hunter with his belt.

COLLABORATE

Your Turn

☑ Identify the **figurative language** Kayla used.

☑ Identify the possessive pronouns that she used in her writing.

☑ Tell how the revisions improved Kayla's writing.

Go Digital!
Write online in Writer's Workspace

289

Essential Question
How do writers look at success in different ways?

Go Digital!

Reaching FOR SUCCESS

A Little League team winning a championship is one kind of achievement. A person petting a dog may not look like an achievement. However, if that person has been scared of dogs his or her whole life, it is a huge achievement.

▶ Do you think success is always a positive thing? Why or why not?

▶ What are some stories that you can think of where the character attains some kind of success?

Talk About It

Write words that describe what you think about success. Then talk with a partner about how you define success.

Success

Vocabulary

Use the picture and the sentences to talk with a partner about each word.

attain

The climber wanted to **attain** the goal of being the first person to reach the peak.

What goal would you like to attain?

dangling

The ripe apple was **dangling** from the end of the branch.

What are other fruits that you might find dangling?

hovering

The hummingbird was **hovering** in front of the flower's petals.

What might a helicopter be hovering over?

triumph

Winning the state soccer championship was a **triumph**!

What is a synonym for triumph?

Poetry Terms

stanza

A **stanza** is two or more lines of poetry that together form a unit of the poem.

Explain how you know when a stanza ends.

denotation

The **denotation** is the basic definition of a word.

What is the denotation of the word little?

connotation

The **connotation** of a word is a meaning suggested by a word in addition to its literal meaning.

What is the connotation of the word scrawny?

repetition

Poets who repeat words or phrases in a poem are using **repetition**.

How might repetition add to a poem's meaning?

Your Turn COLLABORATE

Pick three words. Write three questions for your partner to answer.

Sing to Me

? **Essential Question**

How do writers look at success in different ways?

Read about how two poets share stories of success.

The cool white keys stretched for miles.
How would my hands pull
and sort through the notes,
blending them into music?

I practiced
and practiced all day.
My fingers reaching for a melody
that hung dangling,
like an apple just out of reach.

I can't do this.
I can't do this.

The day ground on,
notes leaping hopefully into the air,
hovering briefly, only to crash,
an awkward jangle, a tangle of noise
before slowly fading away.

My mom found me, forehead on the keys.
She asked, "Would you like some help?
It took months for my hands to do what I wanted."
She sat down on the bench,
her slender fingers plucking notes
from the air.

I can do this.
I can do this.

She sat with me every night that week,
working my fingers until their efforts
made the keys sing to me, too.

— **Will Meyers**

The Climb

"Go on, I dare you!" My brother's voice
mocking, a jaybird's repetitive screech.
We are waiting for the bus
under our immense oak tree.

I reach for the lowest branch and find
another to pull myself up before
I lose my grip on the slippery bark
and slither down the trunk. Again.

Today, at school,
I drop my milk at lunch,
take a pop quiz,
and argue with my friends.

Today is my birthday.
When I get off the bus,
The oak tree doesn't look
any smaller or bigger.

Today, I am ten years old.
I reach for the lowest branch
and find another to pull myself up.
My hands find another and another.

Over and over among the red
outstretched leaves,
foot to branch: push!
hand to branch: pull!

My brother is rooted on the ground,
staring up at me,
until finally, I can't climb any higher,
or I will be a cloud.

— **Sonya Mera**

Make Connections

? Talk about how each poet writes about success. **ESSENTIAL QUESTION**

Compare how the characters in each poem feel to how you feel when you are successful. **TEXT TO SELF**

Narrative Poem

A Narrative Poem:
- Tells a story and has characters.
- Is about fictional or real events.
- May be written in stanzas.

Find Text Evidence

I can tell that both "Sing to Me" and "The Climb" are narrative poems because they both tell a story and have characters.

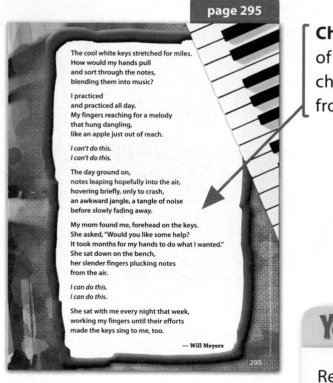

page 295

The cool white keys stretched for miles.
How would my hands pull
and sort through the notes,
blending them into music?

I practiced
and practiced all day.
My fingers reaching for a melody
that hung dangling,
like an apple just out of reach.

I can't do this.
I can't do this.

The day ground on,
notes leaping hopefully into the air,
hovering briefly, only to crash,
an awkward jangle, a tangle of noise
before slowly fading away.

My mom found me, forehead on the keys.
She asked, "Would you like some help?
It took months for my hands to do what I wanted."
She sat down on the bench,
her slender fingers plucking notes
from the air.

I can do this.
I can do this.

She sat with me every night that week,
working my fingers until their efforts
made the keys sing to me, too.

— Will Meyers

295

Character The narrator of the poem is the main character. We see the events from his point of view.

Your Turn

Reread the poem "The Climb." Identify the elements that tell you it is a narrative poem.

298

Theme

The theme is the main message or lesson in a poem. Identifying key details in a poem can help you determine the theme.

 Find Text Evidence

I'll reread "The Climb" on pages 296–297. I will look at the narrator's words and actions to help me identify the theme.

Detail

I lose my grip on the slippery bark/and slither down the trunk. Again.

Key details help you identify the theme.

↓

Detail

The oak tree doesn't look/any smaller or bigger.

↓

Detail

My hands find another and another.

↓

Theme

Persistence leads to success.

Your Turn COLLABORATE

Reread "Sing to Me" on pages 294–295. Find the key details and list them in the graphic organizer. Use the details to determine the theme of the poem.

Stanza and Repetition

A **stanza** is two or more lines of poetry that together form a unit of the poem. Stanzas can be the same length and have a rhyme scheme, or vary in length and not rhyme.

Repetition is the use of repeated words or phrases in a poem. Poets use repetition for rhythmic effect and emphasis.

Find Text Evidence

Reread the poem "The Climb" on pages 296–297. Identify the stanzas and listen for words or phrases that are repeated.

page 296

The Climb

"Go on, I dare you!" My brother's voice
mocking, a jaybird's repetitive screech.
We are waiting for the bus
under our immense oak tree.

I reach for the lowest branch and find
another to pull myself up before
I lose my grip on the slippery bark
and slither down the trunk. Again.

Today, at school,
I drop my milk at lunch,
take a pop quiz,
and argue with my friends.

Today is my birthday.
When I get off the bus,
The oak tree doesn't look
any smaller or bigger.

Today, I am ten years old.
I reach for the lowest branch
and find another to pull myself up.
My hands find another and another.

296

Stanza Each of these groups of lines is a stanza.

Repetition The poet starts the last three stanzas on this page with, "Today."

Your Turn
COLLABORATE

Reread "Sing to Me." What lines does the poet repeat in this poem? What effect does the repetition have on the poem?

Connotation and Denotation

Connotation is a feeling or idea associated with the word.
Denotation is the dictionary's definition of a word.

Find Text Evidence

When I read "The Climb" I know that, besides having a literal meaning, some words make feelings come to mind. In the first stanza, the word immense *means "huge." Connotations of* immense *might be* overwhelming *and* intimidating.

We are waiting for the bus under our immense oak tree.

 **Your Turn**

COLLABORATE

Find an example of connotation and denotation in "Sing to Me" or "The Climb." Give the connotations of the word and its denotation.

301

ImageZoo/Corbis

Readers to . . .

In a poem, writers use sensory details to describe how something looks, sounds, smells, tastes, or feels. Read the excerpt from the poem, "The Climb" below.

Sensory Language

Identify **sensory language** in "The Climb." How does the language help you picture what is happening?

Expert Model

The Climb

"Go on, I dare you!" My brother's voice
mocking, a jaybird's repetitive screech.
We are waiting for the bus
under our immense oak tree.

I reach for the lowest branch and find
another to pull myself up before
I lose my grip on the slippery bark
and slither down the trunk. Again.

Writers

Jack wrote a poem about the ocean.
Read Jack's revisions to his poem.

Editing Marks

⌐⌐ Switch order.

∧ Add.

⌄ Add a comma

⌿ Take out.

(SP) Check spelling.

≡ Make a capital letter.

Grammar Handbook

Pronouns and Homophones
See page 465.

Student Model

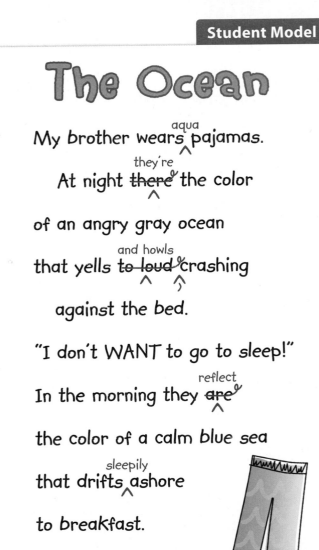

The Ocean

My brother wears ^aqua^ pajamas.

At night ~~there~~ ^they're^ the color

of an angry gray ocean

that yells ~~to loud~~ ^and howls^ crashing

against the bed.

"I don't WANT to go to sleep!"

In the morning they ~~are~~ ^reflect^

the color of a calm blue sea

that drifts ^sleepily^ ashore

to breakfast.

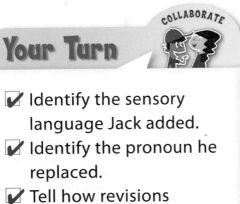

Your Turn

COLLABORATE

☑ Identify the sensory language Jack added.

☑ Identify the pronoun he replaced.

☑ Tell how revisions improved Jack's writing.

Go Digital!
Write online in Writer's Workspace

303

Figure It Out

THE BIG IDEA
What helps you understand the world around you?

PERSEPHONE

Long ago, the ancient Greeks told this myth to help them understand the changing seasons.

Demeter, goddess of the harvest, had a beautiful daughter named Persephone. The joy that Persephone gave Demeter was so great that wherever Demeter stepped, plants, grains, and fruit grew in abundance.

The god Hades looked up from his dark underworld kingdom and fell in love with Persephone. One day, he reached up and dragged Persephone to the underworld. Eventually, he convinced her to marry him.

Demeter searched the whole world for her daughter. She was so broken-hearted that all of the plants began to wither and die around her.

Persephone begged Hades to let her go to her mother. Hades agreed, but before he released her, he gave Persephone a pomegranate. When she ate three of the pomegranate's red seeds, she became bound to the underworld for a part of every year.

When Persephone is with Demeter, everything blooms and grows. But, during the part of the year when she is with Hades, the world becomes a dark, cold place where nothing grows.

Essential Question
In what ways do people show they care about each other?

Go Digital!

Show You Care

People show that they care about each other in different ways. Helping someone with his or her homework, making a special card for a friend, or raking leaves for an elderly neighbor are just some examples of how people show they care about one another.

▶ How do you think the boy in the photo feels about the man in the wheelchair? How can you tell?

▶ What are some ways that you show you care about your friends and family?

Talk About It COLLABORATE

Write words you have learned about how to show you care. Talk with a partner about what you can do to help others.

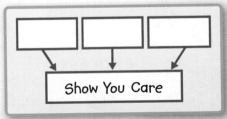

Show You Care

Vocabulary

Use the picture and the sentences to talk with a partner about each word.

bouquet

I assembled the beautiful flowers into a **bouquet**.

How does a bouquet look, smell, and feel?

emotion

Surprise is such a strong **emotion** that people often gasp out loud.

What is an emotion you felt today?

encircle

The children held hands to **encircle** the tree.

How is encircle similar to surround?

express

Tito made a picture to **express** his love of color.

How would you express your love of singing?

fussy

Fussy Mr. Green stood in front of the mirror until his bow tie looked perfect.

What is a synonym for fussy?

portraits

This week Ann's art class is drawing **portraits** of themselves and each other.

What portraits might you see hanging in the White House?

sparkles

The gold beaded curtain **sparkles** in the light.

Name some other things that sparkle.

whirl

The dancers were able to **whirl** and twirl without getting dizzy.

What is a synonym for whirl?

Your Turn

COLLABORATE

Pick three words. Write three questions for your partner to answer.

Go Digital! *Use the online visual glossary*

Sadie's Game

? **Essential Question**

In what ways do people show they care about each other?

Read how a brother shows that he cares about his little sister.

The referee's whistle went off like a shrieking bird, signaling Sadie's second foul of the game. It was only the first quarter, and Sadie had already collided twice with another player's wheelchair. Her coach waved her off the court for a substitution as the crowd shouted catcalls and jeered behind her. She had never seen a crowd **express** such disappointment before.

Sadie watched her teammates **whirl** and spin in front of her. Her **emotions** were all over the place, and it showed in her basketball playing. If only she and her brother had not argued this morning about the game. "What's so important, Richie, that you can't be at the game? Don't I matter anymore?" Sadie had asked.

James Bernardin

Richie was Sadie's whole world, and they both loved sports, especially basketball. Sadie loved to play before her accident, and it was Richie who had taught her to play again afterward. There had been days when she did not want to get out of bed, and he would coax and bully her until she got up. He even borrowed a wheelchair himself to help her learn to play the game all over again. Together they would roll across the outdoor court, zipping, zooming, passing, and dribbling all day long.

But lately Richie preferred to hang out with his new high school friends. Sadie would watch through the window as Richie polished every little nook of his new car. He was as **fussy** as a mother cat cleaning her kittens. When he drove away, Sadie would keep staring out of the window, tears clouding her eyes.

Mama was her sun. Her arms would reach out and **encircle** her in a long, warm embrace. "Sadie," she would say, "your brother loves you. Even though he's got new priorities now, that doesn't mean he doesn't care." But Sadie felt hurt.

Sadie looked up and saw her coach frowning. She searched sadly for her mother, expecting disappointment in her eyes, but instead she saw a wide smile. It was the same happy face she saw in **portraits** of her mother at home. Sadie followed her mother's gaze to find Richie jogging toward her across the gym, holding a purple and white **bouquet** of flowers wrapped tightly with a ribbon. Richie's eyes **sparkled**, and his smile gleamed. He bowed to his sister and handed her the flowers as though she were a queen.

"But we're losing. How do you know we're going to win?" she asked.

"I don't," Richie said. "It's not important. What I know is you're like a whirlwind on the court, and there is no way I am going to miss my little sister's big game!" He put his hand on her shoulder as he said, "It's great to have a lot of new friends, but I realized that you're my best friend."

Sadie smiled. Those words meant more to her than "I'm sorry" ever could. She rested the flowers on her lap and went back out onto the court. Right then Sadie decided to play the rest of the game with the bouquet in her lap. With her brother watching from the sidelines, Sadie stole the ball from an opponent and dribbled her way to the net, making the first of what would be many amazing shots for the team.

Make Connections

Talk about how Richie shows he cares about his little sister, Sadie. **ESSENTIAL QUESTION**

Whom do you care about in the same way that Richie cares about Sadie? Explain how you show you care. **TEXT TO SELF**

James Bernardin

Visualize

When you read, picture the characters, key events, and setting of the story. As you read "Sadie's Game," stop and visualize events to help you better understand the story.

Find Text Evidence

After rereading page 312 of "Sadie's Game," I can use the details to picture the events that are described in the story.

page 312

Richie was Sadie's whole world, and they both loved sports, especially basketball. Sadie loved to play before her accident, and it was Richie who had taught her to play again afterward. There had been days when she did not want to get out of bed, and he would coax and bully her until she got up. He even borrowed a wheelchair himself to help her learn to play the game all over again. Together they would roll across the outdoor court, zipping, zooming, passing, and dribbling all day long.

But lately Richie preferred to hang out with his new high school friends. Sadie would watch through the window as Richie polished every little nook of his new car. He was as **fussy** as a mother cat cleaning her kittens. When he drove away, Sadie would keep staring out of the window, tears clouding her eyes.

Mama was her sun. Her arms would reach out and **encircle** her in a long, warm embrace. "Sadie," she would say, "your brother loves you. Even though he's got new priorities now, that doesn't mean he doesn't care." But Sadie felt hurt.

I visualize Sadie staring out the window as her brother drives off in his car and her mother hugging her. This helps me to understand Sadie's feelings.

Your Turn

COLLABORATE

Reread the last paragraph on page 312 of "Sadie's Game." Visualize how Richie looks as he jogs to Sadie. What words in the text help you picture the scene? As you read, remember to use the strategy Visualize.

Problem and Solution

Identifying the problem and solution in a story can help you understand the characters, setting, and plot. The problem is what the characters want to do, change, or find out. The solution is how the problem is solved.

 Find Text Evidence

As I reread pages 311–312, I can see Sadie has a problem. I will list the key events in the story. Then I can figure out how Sadie finds a solution.

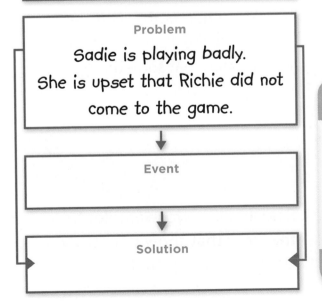

Characters
Sadie, Richie, Sadie's mother

Setting
A basketball court

Problem
Sadie is playing badly. She is upset that Richie did not come to the game.

↓

Event

↓

Solution

Your Turn  COLLABORATE

Reread "Sadie's Game." Find two other important story events. Use these events to identify the solution.

Go Digital!
Use the interactive graphic organizer

315

Realistic Fiction

"Sadie's Game" is realistic fiction.

Realistic fiction:

- Is a made-up story.
- Has characters, settings and events that could exist in real life.
- May contain literary elements such as foreshadowing.

Find Text Evidence

I can tell "Sadie's Game" is realistic fiction. First of all, Sadie is a character who could exist in real life. Also, the setting and events are believable. The story includes foreshadowing.

---

page 312

Richie was Sadie's whole world, and they both loved sports, especially basketball. Sadie loved to play before her accident, and it was Richie who had taught her to play again afterward. There had been days when she did not want to get out of bed, and he would coax and bully her until she got up. He even borrowed a wheelchair himself to help her learn to play the game all over again. Together they would roll across the outdoor court, zipping, zooming, passing, and dribbling all day long.

But lately Richie preferred to hang out with his new high school friends. Sadie would watch through the window as Richie polished every little nook of his new car. He was as **fussy** as a mother cat cleaning her kittens. When he drove away, Sadie would keep staring out of the window, tears clouding her eyes.

Mama was her sun. Her arms would reach out and **encircle** her in a long, warm embrace. "Sadie," she would say, "your brother loves you. Even though he's got new priorities now, that doesn't mean he doesn't care." But Sadie felt hurt.

Sadie looked up and saw her coach frowning. She searched sadly for her mother, expecting disappointment in her eyes, but instead she saw a wide smile. It was the same happy face she saw in **portraits** of her mother at home. Sadie followed her mother's gaze to find Richie jogging toward her across the gym, holding a purple and white **bouquet** of flowers wrapped tightly with a ribbon. Richie's eyes **sparkled**, and his smile gleamed. He bowed to his sister and handed her the flowers as though she were a queen.

312

Foreshadowing Foreshadowing hints at what is going to happen without giving the action away. When Sadie looks up at her mother and sees her smiling, the author is giving the reader a clue that something good is about to happen.

Your Turn

Find and list two examples from the story that show that "Sadie's Game" is realistic fiction.

Similes and Metaphors

A **simile** compares two things by using the words *like* or *as*.
A **metaphor** is the comparison of two things without using *like* or *as*.

 ## Find Text Evidence

I see a simile in the first sentence of "Sadie's Game" on page 311: "The referee's whistle went off like a shrieking bird…." In this sentence, the sound of the whistle is compared to a noisy bird.

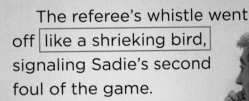

The referee's whistle went off | like a shrieking bird, | signaling Sadie's second foul of the game.

Your Turn

 COLLABORATE

Find the similes and metaphors listed below. Tell what is being compared in each and if it is a simile or a metaphor.

"He was as fussy as a mother cat cleaning her kittens," page 312
"Mama was her sun," page 312
"…as though she were a queen," page 312

James Bernardin

Readers to...

Writers use strong openings, or beginnings, to grab a reader's attention. They do this by using strong verbs and adjectives. Reread the first paragraph from "Sadie's Game" below.

Expert Model

Strong Openings

Identify the adjectives and verbs in the story's **strong opening**. What makes you want to continue reading the story?

The referee's whistle went off like a shrieking bird, signaling Sadie's second foul of the game. It was only the first quarter, and Sadie had already collided twice with another player's wheelchair. Her coach waved her off the court for a substitution as the crowd shouted catcalls and jeered behind her. She had never seen a crowd express such disappointment before.

Writers

Maya wrote an opening to her personal essay. Read Maya's revisions to her opening.

Putting Others First

What does "success" really mean?

~~Success is important.~~ I think

success does not mean having the

most of everything. Selfish people

measure
~~think~~ success by how much money or

how many friends they have. I measure

success by how many people I have

Teaching   younger
helped. ~~Helping~~ my sister ~~learn~~ to ride

green
her ⌐new⌐ shiny bike makes me feel

great. There are many reasons we

should help others succeed, but here

best
is the ~~first~~ one.

Editing Marks

⌐⌐ Switch order.

∧ Add.

∧ Add a comma.

⅁ Take out.

(SP) Check spelling.

≡ Make a capital letter.

Grammar Handbook

Adjectives See page 466.

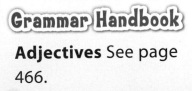

Your Turn

☑ Tell how Maya's opening grabs the reader's attention.

☑ Identify her correct ordering of adjectives.

☑ Explain how Maya's revisions made her opening stronger.

Go Digital!
Write online in Writer's Workspace

Essential Question

What are some reasons people moved west?

Go Digital!

New Beginnings

All through our nation's history, people have moved west. The pioneers moved west for a chance to farm and own land. Prospectors went west to pan for gold. During the Dust Bowl years, people moved west in search of jobs and the hope of a better life.

▶ What might be some other reasons why people decided to move west?

▶ What are some stories that you have read about people moving west?

Talk About It  COLLABORATE

Write some of the words that describe why people moved west. Talk about what people's hopes may have been as they began their journey.

Moving West

321

Vocabulary

Use the picture and the sentences to talk with a partner about each word.

plunging

We watched the dog jump off the rock and go **plunging** into the lake.

What else might go plunging into the water?

prospector

The **prospector** carefully panned for gold in the river.

What are some tools that a prospector might use?

scoffed

Tony's sister **scoffed** at his dream of becoming an Olympic athlete.

What have you scoffed at?

settlement

Pioneers built many **settlements** as they moved west.

What kinds of buildings might you find in a settlement?

shrivel

Grapes left in the hot sun will eventually **shrivel** up and turn into raisins.

What else will shrivel if it is left out in the hot sun?

territories

The pioneers were awed by the size of the **territories** west of the Mississippi.

What did the territories west of the Mississippi eventually become?

topple

The line of dominoes began to **topple** over slowly.

What other things can topple over?

withered

A whole field of sunflowers **withered** and died during the drought.

In what kind of weather might something wither?

COLLABORATE

Your Turn

Pick three words. Write three questions for your partner to answer.

Go Digital! *Use the online visual glossary*

My Big Brother, Johnny Kaw

Essential Question

What are some reasons people moved west?

Read about the Kaw family's journey to settle in Kansas.

Wake up!

I was just a tadpole of a girl when my family decided to leave the crowded city life behind. My daddy said, "There are **territories** out west with wide open spaces. The Kaw family needs room to grow!"

He was mostly talking about my big brother. At fifteen, Johnny had grown so tall that when he stretched out in bed at night his head hung out the front door and his feet hung out the back door all the way into the chicken coop where the hens laid eggs between his toes.

Mama loaded up the wagon with our belongings, and Daddy hitched up the oxen. We began to head west, but it wasn't long until Johnny hollered for everybody to stop.

"We'll never get there with these slowpokes pulling us!" Johnny **scoffed**. He unhitched the team and put one ox on each shoulder.

"Mind you don't let them **topple** off!" Daddy hollered.

"Yes, sir!" Johnny said. "Tadpole can keep an eye on 'em!" He picked me up and set me on top of his head where I had to hang on to handfuls of Johnny's red hair to keep from falling off. Then Johnny grabbed hold of the hitch and began pulling the wagon.

Josee Basaillon

325

He never did have much sense of direction. He pulled that wagon one way then the other, faster and faster, digging out the biggest gully you ever saw. The next night a big rain came and filled it up. I hear that now they call that crooked gully the Kaw River.

Kaw!

Johnny pulled our wagon to a Kansas **settlement** where people were trying to figure out how to raise crops. "Problem is these mountains," one settler said. "They are in the way."

Johnny said that was no problem. He saw a big cottonwood tree, used a saw to cut it down, and whittled it into a giant scythe. Next, he whacked the mountains off down near the ground, hauled them west, and piled them up in a big row. Today folks call them the Rocky Mountains.

Everybody in Kansas was so happy with the nice flat land that they asked us to stay and homestead with them. We built a sod house and started planting wheat.

Now one summer it was mighty dry. All of the wheat had started to **shrivel** up in the field. Our neighbors came and asked for Johnny's assistance. "My crop has about **withered** away to nothing," said one neighbor. "Without rain we're done for!"

"I have got an idea," said Johnny, looking up at some puffy clouds. He grabbed hold of his big hoe and commenced poking holes in the clouds. Down came the rain in buckets, and the wheat was saved!

Josee Basaillon

One morning at the riverbank, Mama was **plunging** our dirty clothes in the water to get them clean when a **prospector** rode up. He said he was headed to California to find gold. "Trouble is," he said, "there's not one decent trail between here and there."

Mama said, "Let me talk to my son."

Johnny was happy to help. For a week he hiked back and forth to all kinds of places dragging his giant bags of wheat everywhere, clearing trails of trees, brush, and boulders. The gold rush folks were tickled to find good clear paths that they named the Oregon Trail, the Santa Fe Trail, and the Chisholm Trail.

I'm sure glad our family ended up in Kansas. Our neighbors tell us that this is a bad place for twisters, but so far we haven't seen one. I can't wait, though! Johnny plans to lasso that twister and ride it like a bucking bronco—and he's promised his little sister a ride!

Make Connections

Talk about why the Kaw family moved to Kansas. **ESSENTIAL QUESTION**

If you could move somewhere new, where would you go? Why? **TEXT TO SELF**

Visualize

When you visualize, you use descriptive details from the story to picture what is happening. As you read "My Big Brother, Johnny Kaw," visualize the characters and key events to help you understand, enjoy, and remember the story.

 Find Text Evidence

In the second paragraph on page 325, I read a description of how tall Johnny Kaw is. This really helps me to visualize him.

> **page 325**
>
> I was just a tadpole of a girl when my family decided to leave the crowded city life behind. My daddy said, "There are **territories** out west with wide open spaces. The Kaw family needs room to grow!"
>
> He was mostly talking about my big brother. At fifteen, Johnny had grown so tall that when he stretched out in bed at night his head hung out the front door and his feet out the back door all the way into the chicken coop where the hens laid eggs between his toes.
>
> Mama loaded up the wagon with our belongings, and Daddy hitched up the oxen. We began to head west, but it wasn't long until Johnny hollered for everybody to stop.
>
> "We'll never get there with these slowpokes pulling us!" Johnny **scoffed**. He unhitched the team and put one ox on each shoulder.
>
> "Mind you don't let them **topple** off!" Daddy hollered.
>
> "Yes, sir!" Johnny said. "Tadpole can keep an eye on 'em!" He picked me up and set me on top of his head where I had to hang on to handfuls of Johnny's red hair to keep from falling off. Then Johnny grabbed hold of the hitch and began pulling the wagon.

I can picture Johnny's head hanging out the front door and his feet out the back door. This helps me understand how big Johnny is.

Your Turn

COLLABORATE

Read page 326 of "My Big Brother, Johnny Kaw." What story events can you visualize? As you read, remember to use the strategy Visualize.

Cause and Effect

A cause is an event or action that makes something happen. An effect is what happens because of the event or action. Identifying the causes and effects in "My Big Brother, Johnny Kaw" can help you understand the sequence of story events.

Find Text Evidence

As I reread page 325 of "My Big Brother, Johnny Kaw," I can look for important cause-and-effect relationships. This will help me better understand the plot of the story.

Cause	➡	Effect
Johnny is extremely tall.	➡	The Kaws decide to move west, for more room.
Johnny thinks the oxen are moving too slowly.	➡	Johnny pulls the wagon himself.
	➡	

Your Turn
COLLABORATE

Reread "My Big Brother, Johnny Kaw." Find more examples of cause-and-effect relationships in the story. Add the information to the graphic organizer.

Go Digital!
Use the interactive graphic organizer

Tall Tale

"My Big Brother, Johnny Kaw" is a tall tale.

A tall tale:

- Is a type of folktale.
- Features a larger-than-life hero.
- Includes hyperbole.

 Find Text Evidence

"My Big Brother, Johnny Kaw" is a tall tale. Johnny Kaw is a larger-than-life character. The story includes examples of hyperbole, such as when the narrator describes her brother.

page 325

Wake up!

I was just a tadpole of a girl when my family decided to leave the crowded city life behind. My daddy said, "There are **territories** out west with wide open spaces. The Kaw family needs room to grow!"

He was mostly talking about my big brother. At fifteen, Johnny had grown so tall that when he stretched out in bed at night his head hung out the front door and his feet hung out the back door all the way into the chicken coop where the hens laid eggs between his toes.

Mama loaded up the wagon with our belongings, and Daddy hitched up the oxen. We began to head west, but it wasn't long until Johnny hollered for everybody to stop.

"We'll never get there with these slowpokes pulling us!" Johnny **scoffed**. He unhitched the team and put one ox on each shoulder.

"Mind you don't let them **topple** off!" Daddy hollered.

"Yes, sir!" Johnny said. "Tadpole can keep an eye on 'em!" He picked me up and set me on top of his head where I had to hang on to handfuls of Johnny's red hair to keep from falling off. Then Johnny grabbed hold of the hitch and began pulling the wagon.

325

Hyperbole is the use of exaggeration for emphasis. The detail that Johnny is so tall that his head hangs out the front door and his feet reach out the back door to the chicken coop emphasizes Johnny's larger-than-life qualities.

Your Turn COLLABORATE

Find and list two more examples that show "My Big Brother, Johnny Kaw" is a tall tale.

Homographs

Homographs are words that are spelled the same but have different meanings and origins. Use context clues to figure out the meanings of the homographs in the story "My Big Brother, Johnny Kaw."

Find Text Evidence

When I read the word head *in the second and third paragraphs on page 325 I can tell it is a homograph. Both words are spelled the same but have different meanings. I will use context clues to figure out the meanings.*

At fifteen, Johnny had grown so tall that when he stretched out in bed at night his head hung out the front door. . . . We began to head west, but it wasn't long until Johnny hollered for everybody to stop.

Your Turn

Use context clues to find the meanings of the following homographs in "My Big Brother, Johnny Kaw."

top, *page 325*
ground, *page 326*
brush, *page 327*

James Bernadine

Readers to...

Writers vary the length of their sentences within a paragraph or story to make the story more interesting. Reread the excerpt from "My Big Brother, Johnny Kaw."

Expert Model

Vary Sentence Types

Identify a **variety of sentence lengths** in the excerpt. How does sentence variety help the sentences flow naturally from one to the next?

One morning at the riverbank, Mama was plunging our dirty clothes in the water to get them clean when a prospector rode up. He said he was headed to California to find gold. "Trouble is," he said, "there's not one decent trail between here and there."

Mama said, "Let me talk to my son."

Johnny was happy to help. For a week he hiked back and forth to all kinds of places dragging his giant bags of wheat everywhere, clearing trails of trees, brush, and boulders. The gold rush folks were tickled to find good clear paths that they named the Oregon Trail, the Santa Fe Trail, and the Chisholm Trail.

Josee Basaillon

Writers

Editing Marks

⊓ Switch order.

∧ Add.

⌃ Add a comma.
⸝

ℒ Take out.

ⓢⓅ Check spelling.

≡ Make a capital letter.

Grammar Handbook

Articles See page 466.

Caleb wrote an essay about tall tales. Read Caleb's revisions to part of his essay.

Student Model

Tall Tales

I love tall tales. They are funny, ~~They~~ *and*

make me laugh. My favorite tall tale is

about Pecos b̲ill because it tells how ~~a~~ *the*

Grand Canyon was created.

Tall tales use lots of ^*exaggerated* detail. For

example, a tall tale about Paul Bunyan

describes him as being so big that he

used a pine tree as ~~an~~ *a* toothpick.

A lot of ~~this~~ *these* tall tales were told

by settlers *moving out west*. I think they helped the

settlers feel less afraid as they camped

out on the wide open plains.

Your Turn COLLABORATE

- ☑ Identify the **varying sentence lengths** that Caleb included.
- ☑ Identify the articles that Caleb used in his writing.
- ☑ Tell how the revisions improved his writing.

Go Digital!
Write online in Writer's Workspace

Essential Question
How can inventions
solve problems?

Go Digital!

S M A R T
SOLUTIONS

Have you ever heard the proverb, "Necessity is the mother of invention"? Inventions are created to solve problems. For example, the rising price of gas has led auto makers to develop smaller, more efficient cars that use less gasoline.

▶ What is a problem that you would like to see solved?

▶ What kind of invention might solve this problem?

Talk About It

Write words that describe how inventions solve problems. Then talk to a partner about an invention that you admire.

Inventions

Vocabulary

Use the picture and the sentences to talk with a partner about each word.

dizzy

Noah felt **dizzy** after spinning around and around on the grass.

What are some things that make you feel dizzy?

experiment

Tony did an **experiment** in class to determine the acidity of a certain liquid.

Why might scientists do experiments?

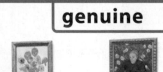

genuine

Are these two paintings in the museum **genuine** or fakes?

What is a synonym for genuine?

hilarious

The **hilarious** movie made the audience laugh nonstop.

What is an antonym for hilarious?

mischief

The dog got into **mischief** and chewed apart the pillow from the couch.

What kind of mischief might a cat get into with a ball of yarn?

nowadays

Nowadays, many people drive smaller cars to conserve gas.

Explain why nowadays many people do not use pay phones.

politician

The **politician** is hoping that the voters will elect her to the state senate.

Why are voters' opinions important to a politician?

procedure

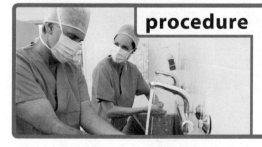

The surgeons followed the hospital **procedure** for sterilizing their hands.

Why is it important to follow the steps in a procedure?

COLLABORATE

Your Turn

Pick three words. Write three questions for your partner to answer.

Go Digital! *Use the online visual glossary*

(t) Radius Images/Alamy; (ct) imagebroker/Alamy; (cb) Exactostock/SuperStock; (b) Chris Ryan/OJO Images/Getty Images

Stephanie Kwolek:
INVENTOR

? Essential Question

How can inventions solve problems?

Read how Stephanie Kwolek invented a super strong fiber that saves lives.

Kevlar® is used in protective vests for police and police dogs.

If you could invent a material for a superhero, what would it be like? It would have to be light, strong, bullet-resistant, and fireproof, right? Chemist Stephanie Kwolek actually invented a material just like this. It's called Kevlar®. Superheroes don't wear it, but everyday heroes like police officers and firefighters do.

Becoming a Chemist

From the time she was young, Stephanie was interested in math and science. She was not the kind of student who caused **mischief**, and she worked hard in school. Stephanie's teachers spotted her talent and talked to her about careers in science. With their encouragement, Stephanie studied chemistry in college. She had hoped to go on to medical school but could not afford it.

Consequently, Stephanie took a job working at a textile lab. She planned to save up enough money from her job so that she could pay for medical school. At the lab, she discovered that she had a **genuine** love of chemistry. She learned how to make chain-like molecules called polymers that could be spun into fabrics and plastics.

Stephanie enjoyed doing **experiments** so much that she decided not to go to medical school.

339

A Strange Liquid

In 1964, Stephanie's lab supervisor asked her to work on making a strong, stiff fiber. The United States was facing a possible gas shortage, and scientists wanted to help. They believed that if you could reinforce tires with a lightweight fiber rather than heavy steel wire, cars and airplanes would use less gasoline. Stephanie began experimenting by mixing polymers. One day, she made an unusual solution, or mixture. Polymer solutions are often thick like molasses. However, this solution was cloudy and watery.

Stephanie brought her strange liquid to the worker in charge of spinning liquids into fibers. He looked at Stephanie's solution and laughed. He thought it was **hilarious** that she believed it could be made into fiber. It looked too much like water and might even clog the spinning machine. But Stephanie kept urging him to spin it until he finally agreed. When he followed the **procedure**, a strong fiber began to form. Stephanie's head spun, and she felt **dizzy** with excitement.

A TIMELINE OF ACHIEVEMENTS

1923	1946	1964	1971	1995
Born in New Kensington, Pennsylvania	Earned a degree in chemistry from Carnegie Mellon University	Discovered the fibers for Kevlar®	Kevlar® first marketed	Inducted into the Inventor's Hall of Fame

Kevlar® is used in cars, such as this solar racing car.

Stronger than Steel

Stephanie tested the fiber in the lab and found that it was fireproof. It was stronger and lighter than steel, too. With these qualities, she believed that the fiber could be turned into a useful material. She was right. The material became known as Kevlar®.

Firefighters wear suits made from Kevlar®.

After Stephanie's discovery, it took almost a decade of teamwork to develop Kevlar®. Some people spent hours on the telephone with the patent office. Others had to think of ways to use and sell it. **Nowadays**, Kevlar® is used by almost everyone. The President and other **politicians** wear protective clothing made from it. So do lumberjacks, firefighters, and police officers. Kevlar® is also used in tires, bicycles, spacecraft, and skis. By developing Kevlar®, Stephanie had found a way to make protective clothing and equipment that is both light and strong.

Stephanie's invention has saved many lives over the years. She was inducted into the National Inventors Hall of Fame for her work, and her photograph has appeared on a book cover and in advertisements for Kevlar®. She says that she never expected to be an inventor but is delighted that her work has helped so many people.

Make Connections

What problems did Stephanie's invention solve? **ESSENTIAL QUESTION**

What would you make out of Kevlar®? Explain why. **TEXT TO SELF**

Bernard Annebique/Sygma/Corbis

Summarize

When you summarize, you retell the most important details in a paragraph or section. To summarize, first identify the key details and then retell them in your own words. Reread "Stephanie Kwolek: Inventor" and summarize sections of the text to make sure you understand them.

 Find Text Evidence

Reread "A Strange Liquid" on page 340. Summarize the most important details in the section.

> **page 340**
>
> **A Strange Liquid**
>
> In 1964, Stephanie's lab supervisor asked her to work on making a strong, stiff fiber. The United States was facing a possible gas shortage, and scientists wanted to help. They believed that if you could reinforce tires with a lightweight fiber rather than heavy steel wire, cars and airplanes would use less gasoline. Stephanie began experimenting by mixing polymers. One day, she made an unusual solution, or mixture. Polymer solutions are often thick like molasses. However, this solution was cloudy and watery.
>
> Stephanie brought her strange liquid to the worker in charge of spinning liquids into fibers. He looked at Stephanie's solution and laughed. He thought it was **hilarious** that she believed it could be made into fiber. It looked too much like water and might even clog the spinning machine. But Stephanie kept urging him to spin it until he finally agreed. When he followed the **procedure**, a strong fiber began to form. Stephanie's head spun, and she felt **dizzy** with excitement.

Kwolek was asked to make a strong fiber. She ended up making an unusual watery polymer solution, which was then spun into a strong fiber.

Your Turn

COLLABORATE

Reread the section "Stronger than Steel" and summarize why Kevlar® is useful. As you read other selections, remember to use the strategy Summarize.

Problem and Solution

Authors use text structure to organize information in a nonfiction text. Problem and solution is one kind of text structure. It presents a problem and then explains the steps taken to solve the problem.

 Find Text Evidence

As I reread "Stephanie Kwolek: Inventor," I will identify problems and the actions taken to solve them. I will also look for words that signal a solution such as consequently *and* as a result.

Problem	Solution
Stephanie Kwolek can't afford to go to medical school.	Consequently, she worked at a textile lab to make money.
A co-worker didn't want to spin the solution into fiber.	Stephanie convinced him to do it.

Your Turn COLLABORATE

Reread "Stephanie Kwolek: Inventor." Look for other problems and solutions. List them in the graphic organizer.

Go Digital!
Use the interactive graphic organizer

343

Biography

"Stephanie Kwolek: Inventor" is a biography.

A biography:

- Is the true story of a real person's life written by another person.
- Usually presents events in chronological order.
- May include text features.

Find Text Evidence

"Stephanie Kwolek: Inventor" is a biography. The key events in Stephanie's life are presented in the order that they happened. There are text features such as a time line and photographs.

page 340

A Strange Liquid

In 1964, Stephanie's lab supervisor asked her to work on making a strong, stiff fiber. The United States was facing a possible gas shortage, and scientists wanted to help. They believed that if you could reinforce tires with a lightweight fiber rather than heavy steel wire, cars and airplanes would use less gasoline. Stephanie began experimenting by mixing polymers. One day, she made an unusual solution, or mixture. Polymer solutions are often thick like molasses. However, this solution was cloudy and watery.

Stephanie brought her strange liquid to the worker in charge of spinning liquids into fibers. He looked at Stephanie's solution and laughed. He thought it was **hilarious** that she believed it could be made into fiber. It looked too much like water and might even clog the spinning machine. But Stephanie kept urging him to spin it until he finally agreed. When he followed the **procedure**, a strong fiber began to form. Stephanie's head spun, and she felt **dizzy** with excitement.

A TIMELINE OF ACHIEVEMENTS

1923	1946	1964	1971	1995
Born in New Kensington, Pennsylvania	Earned a degree in chemistry from Carnegie Mellon University	Discovered the fibers for Kevlar®	Kevlar® first marketed	Inducted into the Inventor's Hall of Fame

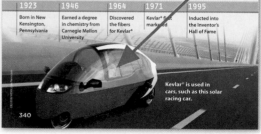

Kevlar® is used in cars, such as this solar racing car.

340

Text Features

Time Line Time lines show events in the order in which they took place.

Photographs and Captions Photographs help you picture information in the text. Captions provide more information.

Your Turn

List two text features that show that "Stephanie Kwolek: Inventor" is a biography. Explain what you learned from the features.

Greek Roots

Knowing Greek roots can help you figure out the meanings of unfamiliar words. Look for words with these Greek roots as you read "Stephanie Kwolek: Inventor."

cycl = circular   *deca = ten*
phot = light   *graph = write*

 Find Text Evidence

When I reread page 341 in "Stephanie Kwolek: Inventor," I see the word telephone. *I know the Greek root* phon *means* sound *and the Greek prefix* tele- *means* far off. *This will help me figure out the meaning of the word.*

Some people spent hours on the telephone with the patent office.

COLLABORATE

Your Turn

Use Greek roots to figure out the meanings of the following words in "Stephanie Kwolek: Inventor."

decade, *page 341*
bicycles, *page 341*
photograph, *page 341*

Readers to...

Writers use transition words or phrases to organize a sequence of events or to move from one idea to another. Reread the paragraph from "Stephanie Kwolek: Inventor" below.

Expert Model

Transitions

Identify **transitions** that connect the sentences. How do these words and phrases connect one idea to the next?

After Stephanie's discovery, it took almost a decade of teamwork to develop Kevlar®. Some people spent hours on the telephone with the patent office. Others had to think of ways to use and sell it. Nowadays, Kevlar® is used by almost everyone. The President and other politicians wear protective clothing made from it. So do lumberjacks, firefighters, and police officers. Kevlar® is also used in tires, bicycles, spacecraft, and skis. By developing Kevlar®, Stephanie had found a way to make protective clothing and equipment that is both light and strong.

Writers

Brady wrote an opinion piece. Read Brady's revisions to one section of his text.

Editing Marks

⎍ Switch order.

∧ Add.

⌄ Add a comma.

✎ Take out.

(SP) Check spelling.

≡ Make a capital letter.

Grammar Handbook

Adjectives That Compare

See page 467.

Student Model

The Internet

There are lots of good inventions, but the internet is the best invention of all. I can't imagine living without ~~it.~~ this ~~It is~~ an amazing invention.

First, the Internet provides tons of information. I use it to do research. Second, the Internet is a great communication tool. I keep in touch ~~on~~ with my cousins using e-mail. Finally, the Internet is important to schools and offices. For example, my school posts homework assignments online.

Your Turn

☑ Identify transition words and phrases that Brady included.

☑ Identify adjectives he used that compare.

☑ Tell how other revisions improved his writing.

Go Digital!
Write online in Writer's Workspace

Essential Question

What can you discover when you look closely at something?

Go Digital!

TAKE A CLOSER LOOK

Look at a peacock feather and you see rings of color. Now look at that same feather under a microscope, and suddenly it resembles a pinecone.

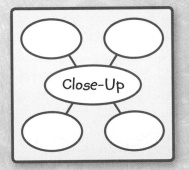

▶ What do you discover when you see an object from far away and then close up?

▶ Why do scientists examine things closely?

▶ What would you like to examine under a microscope?

Talk About It
COLLABORATE

Write words that describe what happens when you look closely at something. Then look at an object and tell what you see.

Close-Up

Vocabulary

Use the picture and the sentences to talk with a partner about each word.

cling

The frog is able to **cling** to the tree branch because of his long legs.

What is a synonym for cling?

dissolves

The tablet **dissolves** quickly in water.

What does sugar look like when it dissolves in water?

gritty

The sand on the bottom of his feet feels **gritty**.

What other things feel gritty?

humid

I like the tropical feeling of the moist, **humid** air in the rain forest.

What is an antonym for humid?

magnify

You can **magnify** a leaf to see its details up close.

How are the words magnify and enlarge similar?

microscope

The scientist used a **microscope** to study the plant cells.

What would you like to view through a microscope?

mingle

The three penguins like to **mingle** and socialize on the beach.

What is an synonym for mingle?

typical

Freezing temperatures in Alaska are normal and **typical** for part of the year.

Describe typical weather for your region.

Your Turn

COLLABORATE

Pick three words. Write three questions for your partner to answer.

Go Digital! *Use the online visual glossary*

Your World Up Close

? ## Essential Question

What can you discover when you look closely at something?

Read about a tool that allows us to see everyday objects up close.

Compare these **gritty** grains of sugar with the magnified sugar crystal.

Does the picture on the left show a diamond or a glass prism? Look closer. Take a step back. You are *too* close.

It is a picture of a sugar crystal. This extreme close-up was taken by an electron microscope, a tool that can **magnify** an item to thousands of times its actual size.

Pictures taken with a high-tech electron microscope are called photomicrographs. The sugar crystal on the left may look huge, but the word *micro* means small. We are seeing a small part of the sugar crystal up close.

Photomicrography dates back to 1840 when a scientist named Alfred Donné first photographed images through a microscope. Around 1852, a German pharmacist made the first version of a camera that took photomicrographs. In 1882, Wilson "Snowflake" Bentley of Vermont became the first person to use a camera with a built-in **microscope** to take pictures of snowflakes. His photographs showed that there is no such thing as a **typical** snowflake. Each is unique. Nowadays, we have electron micrographs.

The photographs of "Snowflake" Bentley showed that snowflakes are shaped like hexagons.

The light microscopes you use in school are weak and do not show much detail. An electron microscope is a much more powerful tool, and it allows scientists to see things we can't see with our own eyes such as skin cells or dust mites.

The picture below is a close-up of human skin and shows the detail an electron microscope can capture. The more an image is magnified, the more detail you will see in the photograph. The most magnification that a photomicrograph can capture is about 2 million times the original image size.

Magnified images have helped scientists to see what causes diseases. Over the years, scientists have learned how these diseases behave. Looking through microscopes, we have even learned what is inside a cell or how a snowflake **dissolves** into a drop of water.

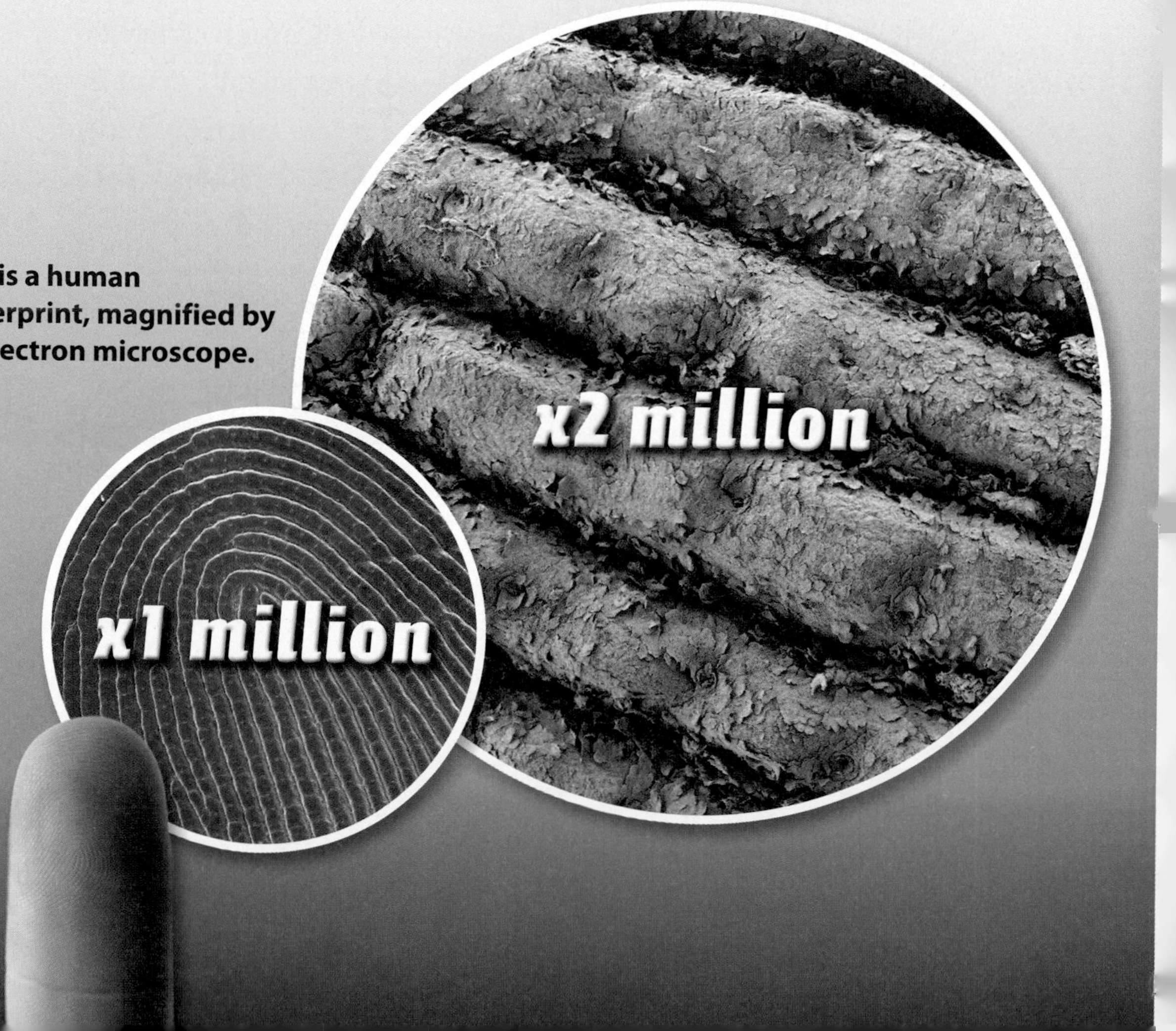

This is a human fingerprint, magnified by an electron microscope.

x2 million

x1 million

When the mold on a strawberry is looked at under an electron microscope, it resembles grapes.

Scientists use electron micrographs to see how objects change over time. For example, we can look at a piece of fruit to see how it decays. First the fruit looks fresh. After a few days it begins to soften. Then specks of mold appear and **cling** to it. Days pass and eventually the fruit is covered in mold. We can see these changes under the microscope far earlier than we can see them with just our eyes.

Suppose you **mingle** outside on a **humid** day with friends. What would the sweat on your skin look like magnified? The possibilities are endless if you examine your world up close.

Make Connections

? How do electron microscopes help scientists? **ESSENTIAL QUESTION**

What objects in your classroom would you like to see under a microscope? **TEXT TO SELF**

Summarize

To summarize a paragraph or a whole selection, retell the key ideas or details in your own words. Reread "Your World Up Close" and summarize sections of the text to make sure you understand them.

Find Text Evidence

Reread the fourth paragraph on page 353. Identify and summarize the key details in the paragraph.

page 353

It is a picture of a sugar crystal. This extreme close-up was taken by an electron microscope, a tool that can **magnify** an item to thousands of times its actual size.

Pictures taken with a high-tech electron microscope are called photomicrographs. The sugar crystal on the left may look huge, but the word *micro* means small. We are seeing a small part of the sugar crystal up close.

Photomicrography dates back to 1840 when a scientist named Alfred Donné first photographed images through a microscope. Around 1852, a German pharmacist made the first version of a camera that took photomicrographs. In 1882, Wilson "Snowflake" Bentley of Vermont became the first person to use a camera with a built-in **microscope** to take pictures of snowflakes. His photographs showed that there is no such thing as a **typical** snowflake. Each is unique. Nowadays, we have electron micrographs.

In 1882, Wilson Bentley was the first person to get close-up pictures of snowflakes. He used a camera attached to a microscope. His photographs showed each snowflake is unique.

Your Turn

COLLABORATE

Reread page 354 of "Your World Up Close" and summarize the key details. As you read, remember to use the strategy Summarize.

Sequence

Authors use text structure to organize information in a nonfiction text. Sequence is one kind of text structure. Authors who use this text structure present information in time order and use words that signal time.

 Find Text Evidence

On page 355 of "Your World Up Close," I read how fruit decays over time. I will look for signal words such as first *and* after.

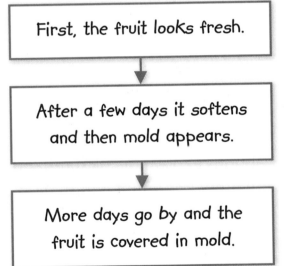

First, the fruit looks fresh.

↓

After a few days it softens and then mold appears.

↓

More days go by and the fruit is covered in mold.

Your Turn

Reread page 353 of "Your World Up Close." Fill in your graphic organizer with details about the development of photomicrography.

Go Digital!
Use the interactive graphic organizer

Expository Text

"Your World Up Close" is an expository text.

Expository text:
- Gives facts and information about a topic.
- Includes text features.

 Find Text Evidence

I can tell that "Your World Up Close" is an expository text. I see photographs and captions. I also see a series of photos that help me to understand the text better.

page 355

When the mold on a strawberry is looked at under an electron microscope, it resembles grapes.

Scientists use electron micrographs to see how objects change over time. For example, we can look at a piece of fruit to see how it decays. First the fruit looks fresh. After a few days it begins to soften. Then specks of mold appear and cling to it. Days pass and eventually the fruit is covered in mold. We can see these changes under the microscope far earlier than we can see them with just our eyes.

Suppose you mingle outside on a humid day with friends. What would the sweat on your skin look like magnified? The possibilities are endless if you examine your world up close.

Make Connections

How do electron microscopes help scientists? **ESSENTIAL QUESTION**

What objects in your classroom would you like to see under a microscope? **TEXT TO SELF**

355

Text Features

Photographs and Captions
Photographs help to illustrate information described in the text. Captions explain the pictures and add other important information about the topic.

 COLLABORATE

Your Turn

Find and list two text features in "Your World Up Close." Tell your partner what information you learned from each of the features.

Antonyms

As you read "Your World Up Close," you may come across a word that you don't know. Sometimes the author will use an **antonym,** another word or phrase that means the opposite of the unfamiliar word.

 Find Text Evidence

On page 353 of "Your World Up Close," I'm not sure what huge *means. I can use the word* small *to help me figure out what* huge *means.*

The sugar crystal on the left may look huge, but the word *micro* means small. We are seeing a small part of the sugar crystal up close.

Your Turn

COLLABORATE

Use context clues and antonyms to find out the meanings of these words in "Your World Up Close."

unique, *page 353*

weak, *page 354*

decays, *page 355*

Readers to ...

Writers use language and a voice that is appropriate for their audience and purpose. When writing an expository text, writers use formal language and a formal voice. Reread the excerpt from "Your World Up Close" below.

Formal Voice

Identify words and phrases that show a **formal voice**. What information do these words and phrases share with the reader?

Expert Model

The most magnification that a photomicrograph can capture is about 2 million times the original image size.

Magnified images have helped scientists to see what causes diseases. Over the years, scientists have learned how these diseases behave. Through microscopes, we have even learned what is inside a cell or how a snowflake dissolves into a drop of water.

x2 million

Writers

Leo wrote about an object. Read Leo's revisions to one section of his essay.

Student Model

MYSTERY OBJECT

When you use this ~~thing~~ ^tool,^ you

should ^make^ sure that the object you want

to magnify is small enough to fit

under the lens. A strand of hair or

a piece of paper ~~could be awesome~~ ^would work well.^

Focusing careful^ly^ on the object will

show ~~most~~ ^more^ detail through the lens.

Some of these tools have powerful

lenses that show the ~~more~~ ^most^ detail of

all. ~~And that's pretty cool.~~ ^And that is an amazing feat.^

Editing Marks

⊓ Switch order.

∧ Add.

∧ Add a comma.
⸜

✍ Take out.

(SP) Check spelling.

≡ Make a capital letter.

Grammar Handbook

Comparing with More and Most
See page 467.

Your Turn

COLLABORATE

☑ Identify examples of formal voice in Leo's essay.

☑ Identify comparisons that correctly use *more* and *most*.

☑ Tell how Leo's revisions improved his writing.

Go Digital!
Write online in Writer's Workspace

Essential Question
How can learning about the past help you understand the present?

Go Digital!

362

TIME FOR KIDS®

TREASURES from the PAST

Archaeologists search for artifacts that will explain how people lived long ago. These clues to the past are a treasure trove of information about the foundations of our country and other countries.

▶ What are some places that archaeologists look for artifacts?

▶ Why is learning about the past important? How can it help us to understand the present?

Talk About It COLLABORATE

Write words that describe why the past is important. Then talk with a partner about a period of history that you are interested in learning about and explain why.

The Past

Vocabulary

**Use the picture and the sentences to talk with a
partner about each word.**

archaeology

The graduate student in **archaeology**
helped to uncover the ancient temple.

How does archaeology help us to learn
about the past?

document

Helen writes in her diary so she can
document the events of her day.

How might an explorer document
her travels?

era

The moon landing in 1969, began a new
era of space exploration.

What invention ended the era of the
horse and buggy?

evidence

The detectives looked for **evidence** at
the crime scene.

Why do detectives look for evidence?

expedition

The wildlife biologist led an **expedition** to explore the rain forest.

What kind of expedition would you like to lead?

permanent

The pyramids were made of large stones so they would stay fixed and **permanent**.

What is a synonym for permanent?

tremendous

I can see a **tremendous** number of stars in the sky tonight.

What is an antonym for tremendous?

uncover

What did you **uncover** when you cleaned the old painting?

What might you uncover if you lift up a big rock by a pond?

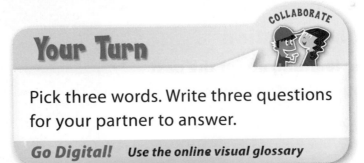

COLLABORATE

Your Turn

Pick three words. Write three questions for your partner to answer.

Go Digital! *Use the online visual glossary*

Where It All Began

Essential Question

How can learning about the past help you understand the present?

Read about the Jamestown settlement.

The building of the Jamestown settlement in 1607

Take a tour of Jamestown, Virginia, the birthplace of America.

They thought they were lost. The *Susan Constant,* the *Godspeed,* and the *Discovery* had sailed from London, England, on December 20, 1606. The **expedition** was bound for Virginia, carrying 144 people.

Finally, on April 26, 1607, the ships sailed into Chesapeake Bay. In the words of one voyager, they found "fair meadows and goodly tall trees." On an island in a river, they built a fort and named it after their king, James. Jamestown would become the first successful, **permanent** English settlement in the New World.

The Struggle to Survive

There is a proverb that says, "Ignorance is bliss." In the case of the 104 men and boys who came ashore, this was true. They were faced with **tremendous** challenges. The water from the James River was not safe to drink, and food was scarce. Two weeks after the settlers arrived, 200 Indians attacked them.

367

John Smith, an experienced military man, became head of the colony in 1608. He had been in charge of finding local tribes willing to swap food for English copper and beads. Smith was tough with both the Indians and Englishmen. "He that will not work, shall not eat," he told the colonists. Smith knew that an attitude of every man for himself would endanger the colony.

Pocahontas saved the life of Captain John Smith.

The western Chesapeake area was ruled by Chief Powhatan, who governed an empire of 14,000 Algonquian-speaking peoples. His daughter Pocahontas became an useful friend and ally to John Smith.

The Real-Life Pocahontas

Princess Matoaka was born around 1595. Her father, Chief Powhatan, called her Pocahontas. She saved John Smith's life twice, and he wrote that Pocahontas's "wit and spirit" were unequaled.

Pocahontas married a planter named John Rolfe, the first marriage in that **era** between an Englishman and a Native American woman. Rolfe, Pocahontas, and their son visited London. She never returned home— she fell ill aboard a ship bound for Jamestown in March 1617 and died.

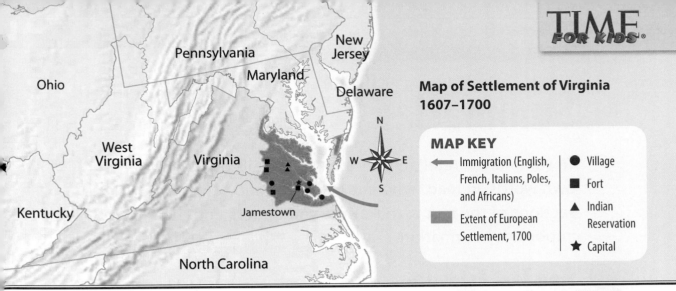

**Map of Settlement of Virginia
1607–1700**

MAP KEY

← Immigration (English, French, Italians, Poles, and Africans)

▬ Extent of European Settlement, 1700

● Village

■ Fort

▲ Indian Reservation

★ Capital

Ohio

Pennsylvania

Maryland

New Jersey

Delaware

West Virginia

Virginia

Kentucky

Jamestown

North Carolina

Taking a Closer Look

Archaeologists digging in Jamestown have discovered Indian artifacts along with English ones, **evidence** that Indians lived in the fort for some time. "It must have been a very close relationship," says William Kelso, an expert in colonial American **archaeology**.

Kelso has worked for 10 years to **document** this site. His team has managed to **uncover** more than 1 million artifacts and has mapped out the fort's shape, its foundations, and a burial ground.

Jamestown left a record of greed and war, but it was also the start of representative government. The settlers gave America a solid foundation to build upon.

Dr. William Kelso working on the archaeological dig in Jamestown

Make Connections

Talk about what archaeologists have found at the Jamestown site. **ESSENTIAL QUESTION**

What would you have liked to ask John Smith about Jamestown? **TEXT TO SELF**

Summarize

To summarize, retell the key ideas or details briefly in your own words. Reread "Where It All Began" and summarize sections of the text to make sure you understand the important information.

Find Text Evidence

Reread the sidebar "The Real-Life Pocahontas" on page 368. Summarize the most important details.

page 368

not work, shall not eat," he told the colonists. Smith knew that an attitude of every man for himself would endanger the colony.

empire of 14,000 Algonquian-speaking peoples. His daughter, Pocahantas became a useful friend and ally to John Smith.

The Real-Life Pocahontas

Princess Matoaka was born around 1595. Her father, Chief Powhatan, called her Pocahontas. She saved John Smith's life twice, and he wrote that Pocahontas's "wit and spirit" were unequaled.

Pocahontas married a planter named John Rolfe, the first marriage in that **era** between an Englishman and a Native American woman. Rolfe, Pocahontas, and their son visited London. She never returned home—she fell ill aboard a ship bound for Jamestown in March 1617 and died.

368

Pocahontas was a famous Native American woman. She was the daughter of Chief Powhatan and saved John Smith's life twice. She was the first Native American woman to marry an Englishman.

Your Turn

COLLABORATE

Reread the section "Taking a Closer Look" on page 369 of "Where It All Began." Summarize the most important details. As you read other selections, remember to use the strategy Summarize.

Sequence

Text structure is the way authors organize and present information in a selection. Sequence is one kind of text structure. Authors present key events in the order in which they happened. Look for dates and words that signal time.

🔍 Find Text Evidence

When I reread page 367 of "Where It All Began," I can look for dates and sequence words such as **finally,** next *and* later, *to understand the order of the events in the text.*

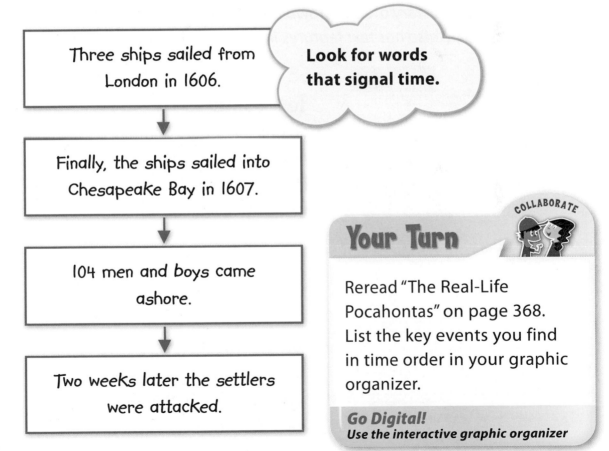

Three ships sailed from London in 1606.

Look for words that signal time.

↓

Finally, the ships sailed into Chesapeake Bay in 1607.

↓

104 men and boys came ashore.

↓

Two weeks later the settlers were attacked.

Your Turn

COLLABORATE

Reread "The Real-Life Pocahontas" on page 368. List the key events you find in time order in your graphic organizer.

Go Digital!
Use the interactive graphic organizer

Informational Article

"Where It All Began" is an informational article.

An informational article:

- Is nonfiction.
- Provides information and facts about people, places, and things.
- May include text features.

 Find Text Evidence

"Where It All Began" is an informational article. It gives facts about the history of Jamestown and the people who lived there. The article also has text features including a map and a sidebar.

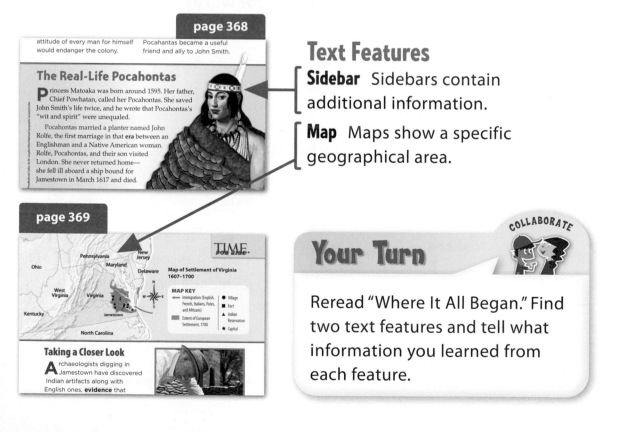

page 368

attitude of every man for himself would endanger the colony.

Pocahontas became a useful friend and ally to John Smith.

The Real-Life Pocahontas

Princess Matoaka was born around 1595. Her father, Chief Powhatan, called her Pocahontas. She saved John Smith's life twice, and he wrote that Pocahontas's "wit and spirit" were unequaled.

Pocahontas married a planter named John Rolfe, the first marriage in that **era** between an Englishman and a Native American woman. Rolfe, Pocahontas, and their son visited London. She never returned home—she fell ill aboard a ship bound for Jamestown in March 1617 and died.

Text Features

Sidebar Sidebars contain additional information.

Map Maps show a specific geographical area.

page 369

Ohio | Pennsylvania | New Jersey
West Virginia | Maryland | Delaware
Kentucky | Virginia | Jamestown
North Carolina

TIME.

Map of Settlement of Virginia 1607–1700

MAP KEY
- Immigration (English, French, Italians, Poles, and Africans)
- Extent of European Settlement, 1700
- Village
- Fort
- Indian Reservation
- Capital

Taking a Closer Look

Archaeologists digging in Jamestown have discovered Indian artifacts along with English ones, **evidence** that

Your Turn COLLABORATE

Reread "Where It All Began." Find two text features and tell what information you learned from each feature.

Proverbs and Adages

Proverbs and adages are short sayings or expressions that have been used for a long time and express a general truth. Every culture has them. Look for context clues to help you figure out the meanings of proverbs and adages.

Find Text Evidence

In the section "The Struggle to Survive" on page 367 of "Where It All Began," I see the proverb ignorance is bliss. *The phrases* tremendous challenges, food was scarce, *and* attacked them *help me to figure out what the proverb means.*

There is a proverb that says, "Ignorance is bliss." In the case of the 104 men and boys who came ashore, this was true. They were faced with tremendous challenges. The water from the James River was not safe to drink, and food was scarce. Two weeks after the settlers arrived, 200 Indians attacked them.

Your Turn

COLLABORATE

Use context clues to determine the meanings of the proverbs and adages below from "Where It All Began."

He that will not work, shall not eat, page 368

Every man for himself, page 368

Readers to . . .

An informational article often ends with a strong conclusion that sums up the main idea of the article. Reread the excerpt from "Where It All Began" below.

Expert Model

Strong Conclusions

Identify the **concluding statement** that sums up the main idea of the article.

Jamestown left a record of greed and war, but it also was the start of representative government. The settlers gave America a solid foundation to build upon.

Writers

Ellen wrote an article about Sybil Ludington. Read Ellen's revisions to a section of her article.

Editing Marks

⊔ Switch order.

∧ Add.

⌄ Add a comma.

✗ Take out.

sp Check spelling.

≡ Make a capital letter.

Grammar Handbook

Comparing with *Good* and *Bad*
See page 467.

Student Model

Sybil Ludington, Patriot

You know all about Paul Revere, but have you ever (sp)herd of Sybil Ludington? She was only 16 years old, but she did the same thing and rode twice as far! ≡many people thought it was not good∧ *bad* for a girl to ride alone at night, but Sybil did a good thing that night and save∧*d* many lives. Sybil Ludington was a true ≡american patriot!

Your Turn   COLLABORATE

☑ Identify the strong concluding sentence.
☑ Identify where Ellen used *good* and *bad*.
☑ Tell how other revisions improved her writing.

Go Digital!
Write online in Writer's Workspace

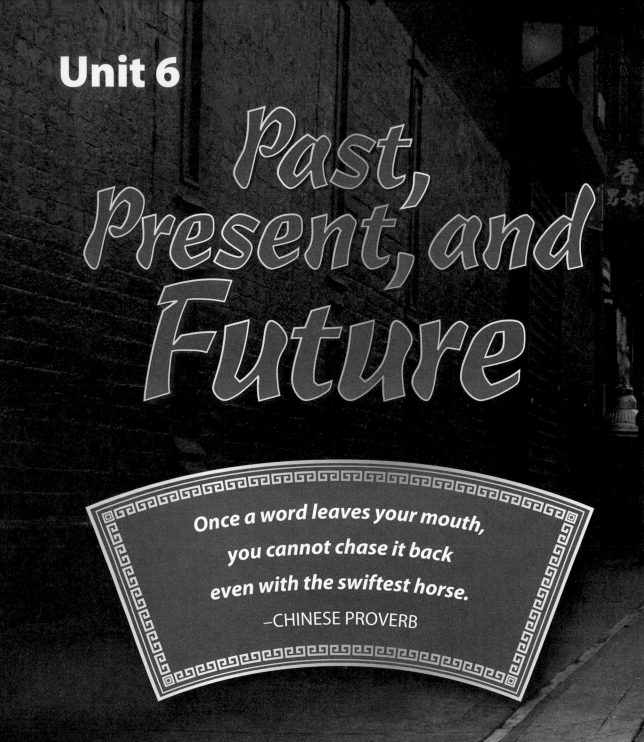

Unit 6

Past, Present, and Future

Once a word leaves your mouth,
you cannot chase it back
even with the swiftest horse.

–CHINESE PROVERB

The Big Idea

How can you build on what came before?

Essential Question

How do traditions connect people?

Go Digital!

SHARING TRADITIONS

Stories, music, and dance are all part of a person's cultural tradition and history. Cultures preserve their traditions by teaching them to the next generation. Keeping cultural traditions alive helps to connect the past to the present.

▶ What do you think the man in the photo is doing?

▶ Why is it important to preserve our traditions?

▶ What are some traditions that you enjoy?

Talk About It COLLABORATE

Write words that describe different traditions. Then talk to your partner about your favorite tradition.

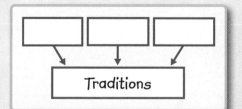

Traditions

Vocabulary

Use the picture and the sentences to talk with a partner about each word.

ancestors

My grandmother showed me a picture of my great-grandparents and other **ancestors**.

Who are some of your ancestors?

despised

Tony liked most vegetables, but he **despised** carrots.

What is an antonym for despised?

endurance

The wheelchair racers had the strength and **endurance** to finish the 20 mile race.

Why is it important for a marathon runner to have endurance?

forfeit

The team had to **forfeit** the game when six players failed to show up.

What is a synonym for forfeit?

honor

One way we **honor** our flag and country is to say the Pledge of Allegiance.

On Veterans Day what are some ways that we honor our veterans?

intensity

The lion roared loudly and with great **intensity**.

Describe a time when you did an activity with intensity.

irritating

Sofia found the loud buzzing of the alarm very **irritating**.

What are some things that you might describe as irritating?

retreated

The turtle **retreated** back into its shell when it sensed danger.

What is an antonym for retreated?

Your Turn

COLLABORATE

Pick three words. Write three questions for your partner to answer.

Go Digital! *Use the online visual glossary*

A Surprise Reunion

Essential Question

How do traditions connect people?

Read how a brother and sister are reunited after many years apart.

Chief Cameahwait looked with **intensity** across the Shoshone camp. The tribe prepared for the Rabbit Dance as warriors oiled their leather costumes. The dance was done to **honor** the rabbit as an important food source. The Shoshone had used traditions such as this dance since the beginning of time to mark special occasions and remember their **ancestors**.

In the distance laughing children were playing with a ball made from rawhide. They rolled the ball into a circle drawn in the dust. If the ball rolled outside the circle, the child must **forfeit** his or her turn. Cameahwait smiled as he remembered the games he had played as a child.

But Cameahwait grimaced beneath his smile. He felt a dull pain in his stomach for his little sister. She had been snatched from the camp during a raid long ago. He **despised** those who had taken her. He closed his eyes and pictured the games they had played together. She had been scrawny and demanding and had an **irritating** habit of following him everywhere, he remembered. He missed her assertive manner and her constant questions. What had become of her?

"It is time to ride," Hawk-That-Soars said, interrupting his thoughts. Cameahwait came back to reality, turned, and mounted his horse.

A man named Captain Lewis had approached the Shoshone days before. Cameahwait knew that Lewis had come in peace, and so he welcomed him and his party. Lewis told the Shoshone his story. He explained that he was part of a company with a mission: he was to explore the land that stretched from the Missouri River to the great ocean. He then asked the chief for a favor. He explained that the rest of his party was waiting at the river with a supply boat. Lewis needed the strength and **endurance** of the Shoshone horses to help transport the supplies across the difficult land. In return Lewis offered the Shoshone food and other goods.

Cameahwait's party arrived at Lewis's camp. There he met Captain Clark.

"Let's sit and discuss how we may help each other," said Clark. He led the men inside a large tent. Buffalo blankets were spread all around. As they settled inside, Lewis addressed the chief. "We travel with a woman who knows your language."

A slender woman with long, dark braids entered the tent. Her eyes adjusted to the dim light filtered through the thick cloth. She nodded to the chief. "I am Sacagawea," she said.

David McCall Johnston

Cameahwait could not believe his eyes! He examined the features of her face. He watched as her expression slowly changed. He immediately knew this was the same sweet face of his lost sister.

Sacagawea quickly ran to him. Tears filled her dark eyes. The pain and sadness that Cameahwait had carried over the years **retreated** to a forgotten place.

"My brother!" she cried. "Is it really you? How long has it been?"

Lewis and Clark were happy to have been unwitting partners in this reunion. Chief Cameahwait promised them he would provide whatever help and resources they needed.

"You have given me a great gift," Cameahwait told them. "You have reunited me with my beloved sister. Our people will sing and tell stories so that all may remember and honor this day for generations to come."

Make Connections

How do traditions and the past connect the chief and his sister? **ESSENTIAL QUESTION**

What traditions do you honor in your family? **TEXT TO SELF**

Reread

When you read historical fiction, you may come across new information or unfamiliar ideas. As you read "A Surprise Reunion," stop and reread any difficult sections of the text to make sure you understand them and remember key details.

Find Text Evidence

You may not be sure what a Rabbit Dance is and why it is part of the Shoshone culture. Reread the first paragraph of "A Surprise Reunion" on page 383.

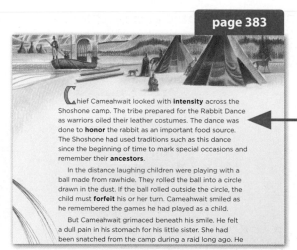

page 383

Chief Cameahwait looked with **intensity** across the Shoshone camp. The tribe prepared for the Rabbit Dance as warriors oiled their leather costumes. The dance was done to **honor** the rabbit as an important food source. The Shoshone had used traditions such as this dance since the beginning of time to mark special occasions and remember their **ancestors**.

In the distance laughing children were playing with a ball made from rawhide. They rolled the ball into a circle drawn in the dust. If the ball rolled outside the circle, the child must **forfeit** his or her turn. Cameahwait smiled as he remembered the games he had played as a child.

But Cameahwait grimaced beneath his smile. He felt a dull pain in his stomach for his little sister. She had been snatched from the camp during a raid long ago. He

I read that the Rabbit Dance is a Shoshone tradition that honors the rabbit as an important food source. I can infer from this that the Shoshone have a close connection to nature.

Your Turn

COLLABORATE

What do Lewis and Clark need from Chief Cameahwait? Reread "A Surprise Reunion" to find out. As you read, remember to use the strategy Reread.

Theme

The theme of a story is the overall message or lesson that an author wants to communicate. To identify the theme, think about what the characters do and say and how they change.

 Find Text Evidence

When I read page 383, I learn that Chief Cameahwait is thinking about his younger sister who was snatched in a raid. He misses her. I think these details are clues to the story's theme.

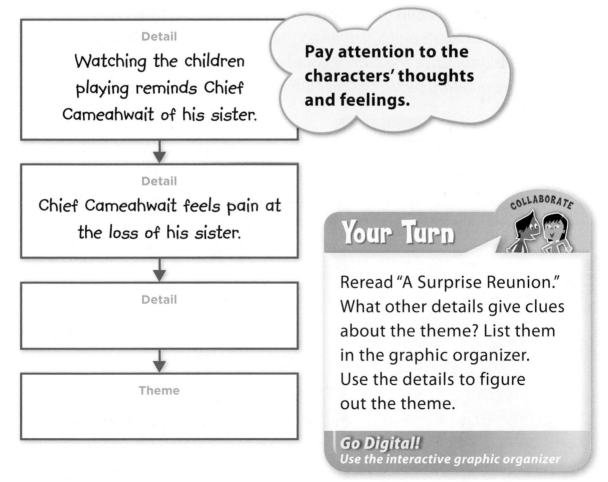

Detail

Watching the children playing reminds Chief Cameahwait of his sister.

> **Pay attention to the characters' thoughts and feelings.**

Detail

Chief Cameahwait feels pain at the loss of his sister.

Detail

Theme

Your Turn

COLLABORATE

Reread "A Surprise Reunion." What other details give clues about the theme? List them in the graphic organizer. Use the details to figure out the theme.

Go Digital!
Use the interactive graphic organizer

Historical Fiction

The selection "A Surprise Reunion" is historical fiction.

Historical fiction:

- Takes place in the past.
- Includes realistic characters, events, and settings.
- May include real people and actual events.
- Includes dialogue.

Find Text Evidence

"A Surprise Reunion" is historical fiction. I know that Chief Cameahwait and Sacagawea are real people. The dialogue is fictional since the author could not know what was said during the meeting between Chief Cameahwait and Sacagawea.

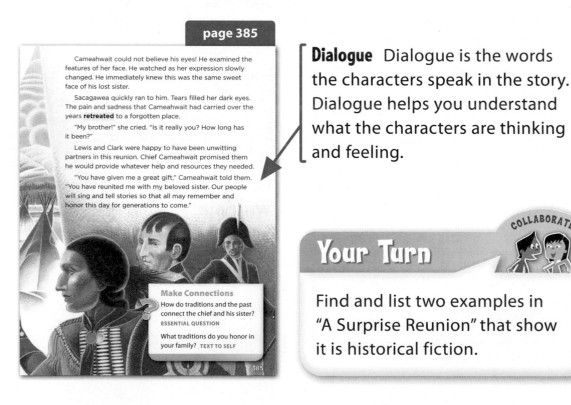

page 385

Cameahwait could not believe his eyes! He examined the features of her face. He watched as her expression slowly changed. He immediately knew this was the same sweet face of his lost sister.

Sacagawea quickly ran to him. Tears filled her dark eyes. The pain and sadness that Cameahwait had carried over the years **retreated** to a forgotten place.

"My brother!" she cried. "Is it really you? How long has it been?"

Lewis and Clark were happy to have been unwitting partners in this reunion. Chief Cameahwait promised them he would provide whatever help and resources they needed.

"You have given me a great gift," Cameahwait told them. "You have reunited me with my beloved sister. Our people will sing and tell stories so that all may remember and honor this day for generations to come."

Make Connections

How do traditions and the past connect the chief and his sister? **ESSENTIAL QUESTION**

What traditions do you honor in your family? **TEXT TO SELF**

385

Dialogue Dialogue is the words the characters speak in the story. Dialogue helps you understand what the characters are thinking and feeling.

COLLABORATE

Your Turn

Find and list two examples in "A Surprise Reunion" that show it is historical fiction.

Connotation and Denotation

Connotation is an idea, meaning, or feeling associated with a word. Denotation is the literal, dictionary definition of a word.

Find Text Evidence

When I read the word scrawny *on page 383 in "A Surprise Reunion," I know its connotation differs from its denotation. The denotation is* very thin. *The connotation is* weak and vulnerable.

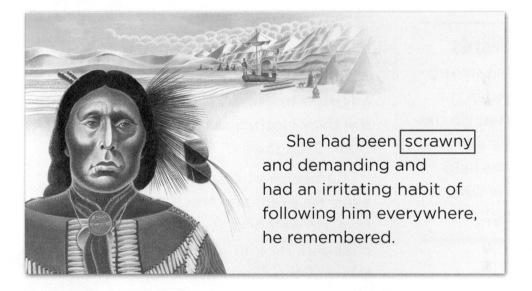

She had been scrawny and demanding and had an irritating habit of following him everywhere, he remembered.

Your Turn

Identify the connotation and denotation of the following words from "A Surprise Reunion."

snatched, *page 383*
assertive, *page 383*
slender, *page 384*

David McCall Johnston

Readers to...

Writers choose strong, descriptive words to make their writing vivid and interesting. Strong words show the action rather than tell and help the reader picture what is happening. Reread the excerpt from "A Surprise Reunion" below.

Expert Model

Strong Words

Identify the **strong words** used in the excerpt. How do the descriptive words and phrases help you picture what is happening?

A slender woman with long, dark braids entered the tent. Her eyes adjusted to the dim light filtered through the thick cloth. She nodded to the chief. "I am Sacagawea," she said.

Cameahwait could not believe his eyes! He examined the features of her face. He watched as her expression slowly changed. He immediately knew this was the same sweet face of his lost sister.

Sacagawea quickly ran to him. Tears filled her dark eyes. The pain and sadness that Cameahwait had carried over the years retreated to a forgotten place.

David McCall Johnston

Writers

Lara wrote about a family tradition. Read Lara's revision of one section of her essay.

Editing Marks

⊓ Switch order.

∧ Add.

⌄ Add a comma.

✐ Take out.

(sp) Check spelling.

≡ Make a capital letter.

Grammar Handbook

Adverbs
See page 468.

Student Model

→ INDEPENDENCE DAY ←

Every year on Independence Day,
~~we~~ watch the fireworks display. In
my family and I go to the park to

fact, everyone in town ~~watches~~ the
gathers to gaze up at

light show. People bring blankets

and happily have a pic(sp)knic under

the stars.

my family always brings a blanket

and some cookies and fruit. The

fireworks reflect ~~beautiful~~ against the
beautifully

lake. It's something I ~~like~~ every year.
look forward to

Your Turn

COLLABORATE

- ✔ Identify strong words and descriptions in Lara's writing.
- ✔ Identify an adverb she included.
- ✔ Tell how the revisions improved Lara's writing.

Go Digital!
Write online in Writer's Workspace

Essential Question

Why is it important to keep a record of the past?

Go Digital!

Untold Stories

For generations, people have come to America to start a new life. We know some of their stories through letters, diaries, and photos that have survived across centuries.

▶ What happens when there is no record of the past?

▶ What do you know about your own family history?

Talk About It

COLLABORATE

Write three words or phrases that describe why keeping a record of the past is important. Then talk to a partner about what you would like somebody to know about you 100 years from now.

Disease Center at Ellis Island

Inspection at Ellis Island

The Past

Vocabulary

Use the picture and the sentences to talk with a partner about each word.

depicts

This painting **depicts** an important moment in our nation's history.

What is a synonym for depicts?

detested

My little sister has always **detested** taking medicine.

What is an antonym for detested?

discarded

The **discarded**, crumpled up paper lay around the trash can.

What is something that you have discarded?

eldest

I am the **eldest** of four children.

What is an antonym for eldest?

ignored

The dog sled team **ignored** the command to stop and kept on running.

Describe a time when you ignored someone or something.

obedience

The dog had no **obedience** training and did not obey his owner's commands.

How are the words obey and obedience related?

refuge

The people took **refuge** in the bus shelter during the rainstorm.

What place do you think of as a refuge?

treacherous

The ice and wet snow made the sidewalks **treacherous** to walk on.

What is something that could be described as treacherous?

COLLABORATE

Your Turn

Pick three words. Write three questions for your partner to answer.

Go Digital! **Use the online visual glossary**

Freedom at FORT MOSE

Essential Question

Why is it important to keep a record of the past?

Read how a boy uses a diary to tell about his life of freedom in a new place.

By September of 1754, twelve-year-old Lucius Jackson and his family had been living at Fort Mose in St. Augustine, Florida, for a year. They were part of a group who had escaped from a plantation in South Carolina. They had heard that Fort Mose was a place of **refuge** for runaways. Over the years many people were willing to endure the **treacherous** journey there in return for the promise of freedom. During his time at Fort Mose, Lucius kept a diary to record what happened there.

17th September 1754

It has been raining for more than a week now. This weather reminds me of my days learning to read and write back in Charleston. When the rains came we couldn't work in the fields, and we were forced to stay in the cabins. We knew that Mr. Slocum, the landowner, detested getting his boots wet so he rarely came to check on us. He thought that all we knew were work and obedience. Miss Celia took a great risk writing letters and words on the dirt floor of the cabin for us children to learn. She said that, as the eldest member in our cabin, it was a risk she was willing to take. Learning to read was easy for me because I was so happy to learn how to turn letters into words and words into ideas. I believe that reading is a gift that cannot be measured. Mr. Samuel Canter believes this, too. He is a farmer who lives near us and who gave me this fine diary. He said, "You are doing a good thing, Lucius. In years to come people can read about this place and understand what we have risked to gain our freedom."

8th October 1754

Last night I got to go on patrol with my father! My duty involved walking along the wall of the fort with him looking and listening for anything unusual. It has been a while since we came under attack, but we cannot let down our guard. We also listen for any people who may be coming here to seek freedom, as we did about one year ago.

While on patrol I thought about the night my family came to Fort Mose and how scared but hopeful all of us felt as we entered through the big heavy gates.

I must stop writing now as it is my turn today to help gather palm fronds, which we lay out in the sun to dry. Once they are dried, they can be used to repair older huts and to build new ones. Each week more people come to the fort. Our priest, Father de Las Casas, keeps the records, and he tells us that there are almost a hundred people now.

26th October 1754

Last week a new family arrived all the way from Virginia and, like everyone else, they arrived almost starved and weak beyond belief. My mother helped the family by giving them clean clothes to replace the ones they had been wearing, and their old ones were quickly discarded. The day after they arrived, I tried to talk to the boy who is about my age, but he ignored me.

The next day, I tried again to speak to the boy whose name is Will. I showed him this diary and explained that it depicts as accurately as possible our life at Fort Mose and the people who come here. He seemed surprised and asked, "You know how to read and write?"

"Yes," I told him. He looked at me without speaking, but I could see a question in his eyes. "Do you want to learn?" I asked him.

"Is it not dangerous?" he asked quietly, looking around to see if anyone could hear us.

I smiled, remembering how long it took me to understand freedom and what it meant.

"Will," I said to my new friend, "here at Fort Mose, you are free to learn, and I am free to teach you."

We began our lessons right away.

Make Connections

Talk about why diaries like Lucius Jackson's represent an important record of the past. **ESSENTIAL QUESTION**

If you could read a diary from any era in the past, what time period would you choose? Why? **TEXT TO SELF**

Neil Shigley

Reread

When you read historical fiction, you may come across facts and ideas that are new to you. As you read "Freedom at Fort Mose," stop and reread important sections of the text to make sure you understand them and can remember key details.

 Find Text Evidence

You may not understand why Lucius Jackson is at Fort Mose. Reread the introduction on page 397.

page 397

By September of 1754, twelve-year-old Lucius Jackson and his family had been living at Fort Mose in St. Augustine, Florida, for a year. They were part of a group who had escaped from a plantation in South Carolina. They had heard that Fort Mose was a place of **refuge** for runaways. Over the years many people were willing to endure the **treacherous** journey there in return for the promise of freedom. During his time at Fort Mose, Lucius kept a diary to record what happened there.

17th September 1754

It has been raining for more than a week now. This weather reminds me of my days learning to read and write back in Charleston. When the rains came we couldn't work in the fields, and we were forced to stay in the cabins. We knew that Mr. Slocum, the landowner, detested getting his boots wet so he rarely came to check on us. He thought that all we knew were work and obedience. Miss Celia took a great risk writing letters

When I reread, I learn that the story takes place in 1754 at Fort Mose in St. Augustine, Florida— a refuge for runaway slaves. From this I can infer that Lucius and his family came to Fort Mose to escape slavery.

Your Turn

COLLABORATE

Why do Lucius and his father have to go on patrol? Reread page 398 of "Freedom at Fort Mose" to find out. As you read other selections, remember to use the strategy Reread.

Theme

A story's theme is the main message or lesson that the author wants to express to the reader. To identify the theme, pay close attention to the characters' words and actions.

 Find Text Evidence

On page 397, I learn that Lucius has been keeping a diary about life at Fort Mose. As I read the first diary entry, I learn that Lucius feels that reading is a gift. These details are clues to the theme.

Detail

Lucius had to learn to read and write in secret.

↓

Detail

Lucius records events at Fort Mose in his diary.

↓

Detail

Lucius writes: "I believe that reading is a gift that cannot be measured."

↓

Theme

Your Turn COLLABORATE

Reread "Freedom at Fort Mose." What other details give clues about the theme? List them in the graphic organizer. Use the details to figure out the theme.

Go Digital!
Use the interactive graphic organizer

Historical Fiction

The selection "Freedom at Fort Mose" is historical fiction.

Historical fiction:

- Takes place in the past.
- Includes realistic characters, events, and settings.
- Usually includes real people and places and may include events that actually happened.
- Is sometimes told as a series of diary entries.

Find Text Evidence

"Freedom at Fort Mose" is historical fiction. I know Fort Mose is a real place that existed at that time in history. Lucius Jackson is a fictional but realistic character who writes in his diary about his life at Fort Mose.

page 398

8th October 1754

Last night I got to go on patrol with my father! My duty involved walking along the wall of the fort with him looking and listening for anything unusual. It has been a while since we came under attack, but we cannot let down our guard. We also listen for any people who may be coming here to seek freedom, as we did about one year ago.

While on patrol I thought about the night my family came to Fort Mose and how scared but hopeful all of us felt as we entered through the big heavy gates.

I must stop writing now as it is my turn today to help gather palm fronds, which we lay out in the sun to dry. Once they are dried, they can be used to repair older huts and to build new ones. Each week more people come to the fort. Our priest, Father de Las Casas, keeps the records, and he tells us that there are almost a hundred people now.

Diary Entries The story is told through a series of diary entries. The reader sees the events of the story through Lucius's eyes.

Your Turn

Find and list two specific details in "Freedom at Fort Mose" that show you it is historical fiction.

Homophones

Homophones are words that sound alike but are spelled differently and have different meanings. Homophone pairs, such as *their* and *there,* are easily confused. Pay attention to the way a homophone is used to help figure out its meaning.

Find Text Evidence

The word heard *in the first paragraph on page 397 is a homophone. The word* herd *sounds the same but is spelled differently and has a different meaning. I know* heard *means, "learned through hearing" and* herd *means, "a group of animals."*

They had ⬚heard⬚ that Fort Mose was a place of refuge for runaways.

Your Turn

COLLABORATE

Look for the following homophones in "Freedom at Fort Mose." Tell the meaning of the word and then identify the word's homophone, its spelling, and meaning.

knew, *page 397*

weak, *page 399*

their, *page 399*

Neil Shigley

Readers to...

Writers organize story events in a logical way. Sequence words and phrases help readers understand when story events occur. Reread the introduction to "Freedom at Fort Mose" below.

Sequence

Identify the **sequence** words and phrases. How do these words help you to understand the setting of the story?

By September of 1754, twelve-year-old Lucius Jackson and his family had been living at Fort Mose in St. Augustine, Florida, for a year. They were part of a group who had escaped from a plantation in South Carolina. They had heard that Fort Mose was a place of refuge for runaways. Over the years many people were willing to endure the treacherous journey there in return for the promise of freedom. During his time at Fort Mose, Lucius kept a diary to record what happened there.

Neil Shigley

Writers

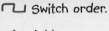

Editing Marks

⊓ Switch order.

∧ Add.

⌃ Add a comma.

✐ Take out.

(SP) Check spelling.

≡ Make a capital letter.

/ Make a lowercase letter.

Ben wrote about his new neighbors. Read Ben's revisions to one section of his essay.

Grammar Handbook

Comparing with Adverbs
See page 469.

Student Model

The New Kid

When someone new moves to

the neighborhood, you should try

                              friendlier

to be ~~more friend like~~ than usual.

First,

Introduce yourself. This always

                              Next,

makes someone feel better! Ask the

person where he or she moved from.

    ask

Then him or her what sports or

hobbies he or she likes best. After

                offer to

talking for a while, show the person

around the neighboorhood.

Your Turn

- ☑ Identify words Ben used that show sequence.
- ☑ Identify examples of adverbs that compare.
- ☑ Tell how other revisions improved Ben's writing.

Go Digital!
Write online in Writer's Workspace

Weekly Concept Resources

Essential Question

How have our energy resources changed over the years?

Go Digital!

gulfimages/Getty Images

ENERGY SOLUTIONS

The building at left has three wind turbines that help produce electricity for the building. Wind energy is one example of a renewable energy source. Oil and gas, or fossil fuels, are nonrenewable energy sources. Once they are used up, they are gone forever.

▶ What is another example of a renewable energy source?

▶ Why is it important to develop new energy sources?

Talk About It

Write words you have learned about energy resources. Talk with a partner about what you can do to help conserve energy.

Energy Resources

Vocabulary

Use the picture and the sentences to talk with a
partner about each word.

coincidence

It was a **coincidence** that Eric bumped
into his friend Tom at the fair.

What kind of coincidence have you
experienced?

consequences

The **consequences** of too much rain can
be flooded roads and fields.

What are some consequences of not
doing your homework?

consume

This kind of car will **consume** less fuel
because it uses less gas than a larger car.

What do people consume?

converted

We **converted** the classroom into a
science lab.

What is a synonym for converted?

efficient

The **efficient** plumber got the job done quickly and easily.

What is an efficient way for you to get to school?

incredible

We saw an **incredible** thunderstorm.

What have you seen that is incredible?

installed

The town **installed** new playground equipment in the park.

What is a synonym for installed?

renewable

When my library card expired, the librarian told me it was **renewable**.

What is something that is not renewable?

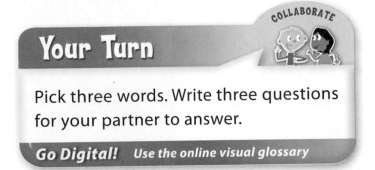

Your Turn

COLLABORATE

Pick three words. Write three questions for your partner to answer.

Go Digital! *Use the online visual glossary*

The Great ENERGY DEBATE

Essential Question

How have our energy resources changed over the years?

Read about a classroom debate over energy resources.

Our energy debate will be an **incredible** event, but I need to study. Our teacher won't tell us which side of the debate we'll be on until the day before it happens, which means we'll have to preplan arguments for both sides.

The debate will be next Tuesday and will include a discussion about different energy sources. Each team will have a microphone. One team will talk about the benefits of an energy source, and the other team will talk about its drawbacks. We'll have to learn about the environmental **consequences** related to each resource, as well as the costs.

411

What Is Energy?

Energy is the ability to do work or make a change. It also is a source of power for making electricity or doing mechanical work. We use the wind, the sun, fossil fuels, and biofuels to produce energy. Burning coal produces heat energy that is converted into electrical energy. We use that energy to light our houses. Solar energy comes from the sun. Solar panels convert sunlight into electrical energy.

We may be asked to debate the future of gasoline as an energy source. If so, I would say that gasoline is made from oil, a fossil fuel. According to geologists, fossil fuels formed over hundreds of millions of years from ancient plant and animal remains. But here's the problem: we use these fuels far faster than it takes them to form. Because fossil fuels are nonrenewable resources, if we keep using them eventually there will be none left. Plus burning these fuels pollutes the air!

It is easy to be hypercritical of fossil fuels. However, most of our cars and factories use this type of fuel, and therefore changing everything would be a huge undertaking.

If we are asked to debate the use of wind energy, we would have to know that this is a **renewable** energy source. For example, unlike fossil fuels, wind will never run out. One large wind turbine could produce enough energy for a whole city! In addition, this method doesn't damage the environment. Turbines can be placed all over the world to capture wind energy. Then the energy from the turbines is **converted** into electrical energy. But there is a drawback. Wind may not be as **efficient** as other energy sources. Only about 30 or 40 percent of all wind energy is changed into electricity. It would be very expensive to have wind turbines **installed** all over the world.

This debate is important for people in the United States. Our country makes up only about 5 percent of the entire world's population. Yet we **consume** about 30 percent of the world's energy. It is not a **coincidence** that students are asked to take part in these debates. We will probably have to make these decisions when we are adults. The debate will be difficult, but I will be ready!

Make Connections

How might our dependence on fossil fuels change in the future? **ESSENTIAL QUESTION**

What can you do to help save energy resources? **TEXT TO SELF**

Ask and Answer Questions

When you read an informational text, you may come across new information. Asking questions about the text and reading to find the answer can help you to understand new information. As you read "The Great Energy Debate," ask and answer questions about the text.

Find Text Evidence

When you first read "The Great Energy Debate," you may have asked yourself why the narrator said on page 411 that the students had to "preplan arguments for both sides."

page 411

Our energy debate will be an **incredible** event, but I need to study. Our teacher won't tell us which side of the debate we'll be on until the day before it happens, which means we'll have to preplan arguments for both sides.

The debate will be next Tuesday and will include a discussion about different energy sources. Each team will have a microphone. One team will talk about the benefits of an energy source, and the other team will talk about its drawbacks. We'll have to learn about the environmental **consequences** related to each resource, as well as the costs.

The text says the teacher wouldn't tell which side of the issue students would be debating. Therefore, I inferred that the students had to study pros and cons for each side.

Your Turn

COLLABORATE

Reread "The Great Energy Debate" to ask and answer questions of your own. As you read, remember to use the strategy Ask and Answer Questions.

Main Idea and Key Details

The main idea is the most important idea or point that an author makes in a paragraph or section of text. Key details give important information to support the main idea.

 Find Text Evidence

When I reread the first paragraph of "The Great Energy Debate" on page 412, I can identify the key details. Next I can think about what those details have in common. Then I can figure out the main idea of the section.

Main Idea

If we keep using fossil fuels, eventually there will be none left.

> **The key details tell about the main idea.**

Detail

Fossil fuels take hundreds of millions of years to form.

Detail

We use fossil fuels faster than it takes them to form.

Detail

Fossil fuels are nonrenewable resources.

Your Turn

Reread the first paragraph on page 413. Find the key details and list them in your graphic organizer. Use the details to determine the main idea.

Go Digital!
Use the interactive graphic organizer

Narrative Nonfiction

"The Great Energy Debate" is narrative nonfiction.

Narrative nonfiction:

- Tells a story.
- Presents facts and information about a topic.
- Includes text features.

Find Text Evidence

I can tell that "The Great Energy Debate" is narrative nonfiction. It tells a story about students preparing for a debate while providing facts about energy sources. It also has text features.

> ### page 412
>
> **What Is Energy?**
> Energy is the ability to do work or make a change. It also is a source of power for making electricity or doing mechanical work. We use the wind, the sun, fossil fuels, and biofuels to produce energy. Burning coal produces heat energy that is converted into electrical energy. We use that energy to light our houses. Solar energy comes from the sun. Solar panels convert sunlight into electrical energy.
>
> We may be asked to debate the future of gasoline as an energy source. If so, I would say that gasoline is made from oil, a fossil fuel. According to geologists, fossil fuels formed over hundreds of millions of years from ancient plant and animal remains. But here's the problem: we use these fuels far faster than it takes them to form. Because fossil fuels are nonrenewable resources, if we keep using them eventually there will be none left. Plus burning these fuels pollutes the air!
>
> It is easy to be hypercritical of fossil fuels. However, most of our cars and factories use this type of fuel, and therefore changing everything would be a huge undertaking.
>
> 412

Text Features

Sidebars Sidebars provide more information to help explain the topic. Sidebars are read after the main part of the text.

Your Turn

COLLABORATE

Find and list two text features in "The Great Energy Debate." Explain what you learned from each feature.

Latin and Greek Prefixes

A prefix is a word part added to the front of a word to change its meaning. Some prefixes come from Latin, such as:

      non- = not   *pre- = before*

Other prefixes come from Greek, such as:

      hyper- = excessively   *bio- = life*

Find Text Evidence

In "The Great Energy Debate," I see the word biofuels *on page 412.* Bio- *is a Greek prefix that means "life." So* biofuels *are fuels that come from living things.*

We use the wind, the sun, fossil fuels, and biofuels to produce energy.

COLLABORATE

Your Turn

Use your knowledge of prefixes and context clues to find the meanings of the words in "The Great Energy Debate."

    preplan, *page 411*
    nonrenewable, *page 412*
    hypercritical, *page 412*

Readers to...

Writers use transition words to organize a sequence of events or to move from one idea to another. Reread the excerpt from "The Great Energy Debate" below.

Expert Model

Transitions

Identify the transitions. How do the **transition words** help the reader move from one idea to another?

If we are asked to debate the use of wind energy, we would have to know that this is a renewable energy source. For example, unlike fossil fuels, wind will never run out. One large wind turbine could produce enough energy for a whole city! In addition, this method doesn't damage the environment. Turbines can be placed all over the world to capture wind energy. Then the energy from the turbines is converted into electrical energy. But there is a drawback. Wind may not be as efficient as other energy sources. Only about 30 or 40 percent of all wind energy is changed into electricity.

Writers

Kim wrote about saving energy. Read Kim's revisions to a section of her essay.

Grammar Handbook

Negatives See page 470.

Student Model

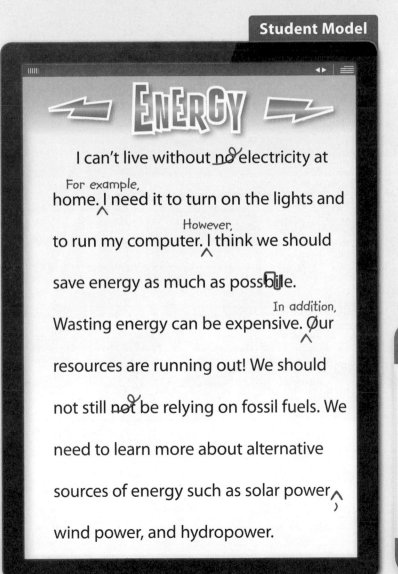

ENERGY

I can't live without ~~no~~ electricity at

For example,
home. I need it to turn on the lights and

However,
to run my computer. I think we should

save energy as much as possble.

In addition,
Wasting energy can be expensive. Our

resources are running out! We should

not still ~~not~~ be relying on fossil fuels. We

need to learn more about alternative

sources of energy such as solar power,

wind power, and hydropower.

Your Turn

COLLABORATE

- ☑ Identify transitions that Kim included.
- ☑ Identify the correct use of a negative.
- ☑ Tell how revisions improved Kim's writing.

Go Digital!
Write online in Writer's Workspace

Essential Question
What has been the role of money over time?

Go Digital!

HERE'S MY DOLLAR!

Do you have a dollar bill? That dollar can be exchanged for any number of things. Centuries ago, people had to barter, or trade, for the goods and services that they wanted.

► What are the different ways that we use money?

► How do you think we will pay for things in the future?

Talk About It COLLABORATE

Write words that tell how money is used. Then discuss what you think money will look like in the future.

Money

Vocabulary

Use the picture and the sentences to talk with a partner about each word.

currency

I exchanged American money for foreign **currency** at the bank.

What kind of currency do we use in the United States?

economics

Bartering is a system of **economics** where people trade one thing for another.

What can you learn about economics by opening a lemonade stand?

entrepreneur

Sarah is an **entrepreneur** who started her own dog walking business.

If you became an entrepreneur, what business would you start?

global

The Internet is a **global** electronic network that connects people around the world.

How is local different from global?

invest

Victoria wants to **invest** more of her allowance in her stamp collection.

What is another thing you might invest money in?

marketplace

Lauren and her mother visited the **marketplace** to buy fresh vegetables.

What else can people buy at a marketplace?

merchandise

The clothing shop's **merchandise** includes dresses, skirts, and tops.

What kind of merchandise is sold in an electronics store?

transaction

The man gave his credit card as part of the **transaction** to pay for his breakfast.

What might someone use as money during a transaction on the Internet?

COLLABORATE

Your Turn

Pick three words. Write three questions for your partner to answer.

Go Digital! **Use the online visual glossary**

THE HISTORY of MONEY

Essential Question

What has been the role of money over time?

Read about the history of money.

A painting of a commercial center in Beijing, China, in 1840

424

What makes money valuable? If you think about it, a dollar bill is only a piece of paper. You cannot eat, wear, or live in a dollar bill. So why do people want it? Think about the proverb, "Money doesn't grow on trees." Money is considered valuable because it is hard to get.

Bartering

Imagine you're a goat herder visiting a **marketplace** in China in 1200 B.C. The **merchandise** being sold around you ranges from cattle to tools. Suppose you need to purchase a piece of rope. How will you pay for it? The goats you own are your sole source of income so you would not want to trade a goat for the rope. The goat is too valuable! Instead, you might trade goat milk for the rope. This system of **economics** is called bartering. But what if the rope merchant does not want goat milk?

Early Currency

No need to cry over spilt milk. Luckily, you sold some goat milk earlier in the day in exchange for ten cowrie shells, the first system of **currency** in China. You hand two cowrie shells to the rope merchant and put the rest in your pocket. This is a much easier way to buy and sell things. Cowrie shells are lightweight, durable, and easier to take with you than a goat. The idea of currency is catching on around the world in Thailand, India, and Africa.

You decide to save your extra shells until you have enough to **invest** in another goat. You will be spending cowries with the expectation that another goat will pay off later since you can drink or sell the milk it produces. Taking this type of business risk makes you an **entrepreneur**.

This painting shows a scene from a typical 19th century Italian market.

New Kinds of Currency

If you were at a marketplace in Rome around 900 B.C., you might have used salt as a form of currency. The idiom "to be worth one's salt" is still used today.

Another form of currency, metal coins, first emerged in China around 1000 B.C. Coins varied in shape, size, and worth. By the 7th century B.C., coins made of precious metals such as silver and gold became popular in Europe and the Middle East. These coins were usually round. After being weighed on a scale to determine their value, coins were stamped with designs that stated their worth.

GLOSSARY OF MONEY TERMS

BARTERING (BAR-tur-ing) Trading by exchanging food, services, or goods instead of using money.

CURRENCY (KUR-uhn-see) Any form of money that is used in a country.

ECONOMY (ee-KON-uh-mee) A system or method of managing the production and distribution of money, goods, and services.

MARKETPLACE (MAR-kit-plays) A place where food and goods are bought and sold, or the world of business, trade, and economics.

Paper Money

Carrying a bag of coins can be heavy. The weight of coins and a metal shortage are two reasons the use of paper money developed in China in the 10th century. The earliest European paper money appeared in Sweden at the beginning of the 17th century. Italy started to use paper money about 90 years later. Paper money originally represented the gold or silver a person had in the bank. Today, we can tell the value of paper money by reading the numbers printed on it.

Modern Money

In today's **global** economy, exchanging money electronically is common. Many people use a credit or debit card to make a digital **transaction**. Numbers on a computer screen represent dollars and cents, but no actual paper money is exchanged.

As easy as it is to spend money today, saving money is important. When considering spending money, think of the famous proverb, "A penny saved is a penny earned."

Make Connections

Why did using currency replace bartering? **ESSENTIAL QUESTION**

How does money affect your daily life? **TEXT TO SELF**

(b) Jose Luis Pelaez-Inc./Blend Images/Getty Images; (t) Dorling Kindersley/Getty Images

Ask and Answer Questions

When you read informational text, you can ask questions before, during, and after reading to help you understand the text and remember the information. As you read "The History of Money," look for answers to your questions.

 Find Text Evidence

You may ask yourself why paper money was an improvement over coins. Reread the section "Paper Money" on page 427 of "The History of Money" to find the answer.

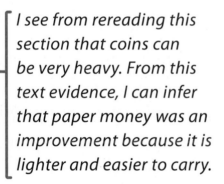

page 427

Paper Money

Carrying a bag of coins can be heavy. The weight of coins and a metal shortage are two reasons the use of paper money developed in China in the 10th century. The earliest European paper money appeared in Sweden at the beginning of the 17th century. Italy started to use paper money about 90 years later. Paper money originally represented the gold or silver a person had in the bank. Today, we can tell the value of paper money by reading the numbers printed on it.

Modern Money

In today's **global** economy, exchanging money electronically is common. Many people use a credit or debit card to make a digital **transaction**. Numbers on a

I see from rereading this section that coins can be very heavy. From this text evidence, I can infer that paper money was an improvement because it is lighter and easier to carry.

Your Turn

COLLABORATE

As you reread "The History of Money" ask your own question and then read to find the answer. As you read other selections, remember to use the strategy Ask and Answer Questions.

Main Idea and Key Details

The main idea is the most important idea or point that an author makes in a paragraph or section of text. Key details give important information to support the main idea.

 Find Text Evidence

When I reread "New Kinds of Currency" on page 426 of "The History of Money," first I can look closely to find the key details. Next I can think about what they have in common. Then I can figure out the main idea of the section.

Main Idea
The first metal coins varied in shape, size, and worth.

Detail
Coins were made of precious metals such as silver and gold.

Detail
Coins were usually round.

Detail
Coins were stamped with designs that stated their worth.

All the key details tell about the main idea.

COLLABORATE

Your Turn

Reread the section "Paper Money" on page 427 of "The History of Money." Find the key details and list them in your graphic organizer. Use the details to determine the main idea of the section.

Go Digital!
Use the interactive graphic organizer

Expository Text

"The History of Money" is an expository text.

Expository text:

- Explains facts and information about a topic.
- Includes text features.

Find Text Evidence

I can tell "The History of Money" is an expository text. It gives me facts and information about the kinds of money that have been used throughout history. It includes a variety of text features.

page 426

This painting shows a scene from a typical 19th century Italian market.

New Kinds of Currency

If you were at a marketplace in Rome around 900 B.C., you might have used salt as a form of currency. The idiom "to be worth one's salt" is still used today.

Another form of currency, metal coins, first emerged in China around 1000 B.C. Coins varied in shape, size, and worth. By the 7th century B.C., coins made of precious metals such as silver and gold became popular in Europe and the Middle East. These coins were usually round. After being weighed on a scale to determine their value, coins were stamped with designs that stated their worth.

426

GLOSSARY OF MONEY TERMS

BARTERING (BAR-tur-ing) Trading by exchanging food, services, or goods instead of using money.

CURRENCY (KUR-uhn-see) Any form of money that is used in a country.

ECONOMY (ee-KON-uh-mee) A system or method of managing the production and distribution of money, goods, and services.

MARKETPLACE (MAR-kit-plays) A place where food and goods are bought and sold, or the world of business, trade, and economics.

Text Features

Headings Headings explain what kind of information is in each section of the text.

Glossary A glossary defines words that are important to the topic of the selection. It lists the words in alphabetical order.

COLLABORATE

Your Turn

Find and list two text features in "The History of Money." Tell what information you learned from each text feature.

Proverbs and Adages

Proverbs and adages are short sayings that have been used for a long time. They usually express a general truth or observation. Every culture has them. Look for context clues to help you figure out the meanings of proverbs and adages.

 Find Text Evidence

When I reread "Modern Money" on page 427 of "The History of Money," I see the proverb, "A penny saved is a penny earned." The previous sentence helps me to figure out what the proverb means.

As easy as it is to spend money today, saving money is important. When considering spending money, think of the famous proverb, "A penny saved is a penny earned."

 **Your Turn**

Use paragraph clues to determine the meanings of the proverbs and adages below from "The History of Money."

Money doesn't grow on trees. page 425

No need to cry over spilt milk. page 425

Readers to ...

Writers use content words to explain a topic. Content words are specific words that relate to a topic. Reread the excerpt from "The History of Money" below.

Expert Model

Content Words

Identify the **content words**. How do the content words help you learn about the topic?

The idea of currency is catching on around the world in Thailand, India, and Africa.

You decide to save your extra cowrie shells until you have enough to invest in another goat. You will be spending cowries with the expectation that another goat will pay off later since you can drink or sell the milk it produces. Taking this type of business risk makes you an entrepreneur.

Writers

Teddy wrote about why people work. Read Teddy's revisions to one section of his text.

Editing Marks

⎡⎤ Switch order.

∧ Add.

∧ Add a comma.

⌷ Take out.

(SP) Check spelling.

☰ Make a capital letter.

Grammar Handbook

Prepositions
See page 471.

Student Model

Why People Work

to earn an income
People work ∧ so that they can

pay for
~~get~~ the things they need and want.

People also work because it keeps

our economy healthy. I think earning

money is important. When I wanted

a new bike ∧ my parents told me I had

to earn it by working. ~~a~~fter walking

for three months,
the neighbor's dog ∧ I had enough

money to buy a new bike.

The next day
∧ I went to buy my new bike from

Al's Bike Shop.

Your Turn

- ✔ Identify content words that Teddy included.
- ✔ Identify a preposition he used.
- ✔ Tell how Teddy's revisions improved his writing.

Go Digital!
Write online in Writer's Workspace

433

Essential Question
What shapes a person's identity?

Go Digital!

Who Am I?

The people in your life help shape who you are, yet you are unique. Think about the people around you, such as your family and friends.

▶ How are you like them? How are you different from them?

▶ What events in your life have influenced you the most?

▶ Where do your ancestors come from? Knowing your roots can help you understand your identity.

Talk About It

Write words that describe who you are. Then talk with a partner about what has helped shape who you are.

My Identity

Corbis Flirt/Alamy

Vocabulary

Use the picture and the sentences to talk with a partner about each word.

gobble

Ted saw the hungry dog **gobble** up his dinner in less than a minute.

What other kind of animal might gobble up its dinner?

individuality

Sara expressed her **individuality** by wearing a unique pair of slippers.

How are the words originality and individuality similar?

mist

The spray from the sprinklers created a wet **mist**.

What things can you spray that produce a mist?

roots

My family has **roots** in California, but most of our family lives in Florida.

What have you learned about your roots?

Poetry Terms

metaphor

A **metaphor** compares two unlike things without the use of *like* or *as*.

Use a metaphor to describe a school bus.

personification

Personification is when human characteristics are given to anything that is not human.

What would be an example of personification?

imagery

Imagery is the use of words to create a picture in the reader's mind.

How would using sensory details help to create imagery in a poem?

free verse

Free verse poems do not have a consistent metrical pattern or rhyme scheme.

Why might a poet choose to write in free verse?

Your Turn

COLLABORATE

Pick three words. Write three questions for your partner to answer.

Go Digital! **Use the online visual glossary**

Essential Question

What shapes a person's identity?

Read how poets talk about important experiences.

CLIMBING BLUE HILL

When the yellow leaves begin to
glimmer among the green ones,
we hike up Blue Hill
through an early morning mist.

"It's not much farther, boys!"
My grandfather bellows happily,
his words an echo of all the other times
he's had to urge us up a steep trail.

I hear the comforting squeak of his boots
as the ground's chill breath whispers
against our ankles and the overgrown
branches tug curiously at my hair.

Abruptly, the trail spits us out,
onto gray rock, into blue sky and sunlight.
My brother shouts, shoves me aside,
races to the low bushes huddled against the wind.

His fingers tug at the tiny leaves.
"Look! Blueberries!" He yells.
And we gobble the blue sweetness up,
my brother, my grandfather, and me.

— **Andrew Feher**

Susan Gal

My Name Is Ivy

"Why did I name you after a plant?

Look, this is ivy," my mother explains,
pointing at an intricate fan

of glossy green heart-shaped leaves
decorating the side
of our house.
"Ivy will grip onto anything,
will grow where it wants to go.
Will use its long skinny fingers
to find a way over
brick walls, up stone walls,
will climb a roof and keep on
going until it touches
the stars."

— Bryce Neale

Collage

Grandma gave me her eyes.
"Eyes of a panther," Grandpa whispers.

Grandpa gave me his nose.
"A bumpy, rocky road of a nose," Grandma scoffs.

Dad gave me his long skinny toes.
"My roots reach back to the lemurs," he jokes.

Mama gave me her lopsided smile.
"Don't ever lose it," she warns.

And I gave them my heart.
"It's big enough to hold you all," I say.

— **Maria Diaz**

Make Connections

? What do these poets think shapes
a person's individuality?
ESSENTIAL QUESTION

What has influenced you? **TEXT TO SELF**

Susan Gal

441

Free Verse

Free Verse:
- Does not have a rhyme scheme or a metrical pattern.
- May have irregular lines.

 Find Text Evidence

I can tell that "My Name Is Ivy" is a free verse poem because it does not have a rhyme scheme or a metrical pattern.

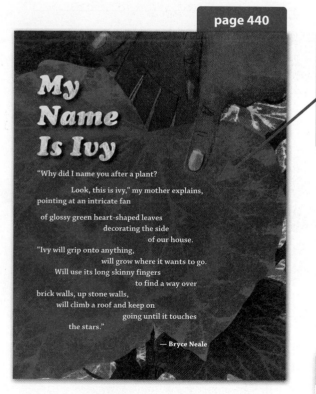

page 440

My Name Is Ivy

"Why did I name you after a plant?

Look, this is ivy," my mother explains,
pointing at an intricate fan

of glossy green heart-shaped leaves
decorating the side
of our house.
"Ivy will grip onto anything,
will grow where it wants to go.
Will use its long skinny fingers
to find a way over
brick walls, up stone walls,
will climb a roof and keep on
going until it touches
the stars."

— Bryce Neale

I wonder why the lines are all spread out. The lines in this poem are not the same length. The poet chose to give the lines a zigzag pattern.

 COLLABORATE

Your Turn

Reread the poem "Collage." Explain why it is a free verse poem.

Theme

The theme is the main message or lesson in a poem. Identifying the key details in a poem can help you determine the theme.

 Find Text Evidence

All of the poems in this lesson are about identity, but each poem has a different theme. I'll reread "Collage" on page 441 and look for key details to determine the theme of the poem.

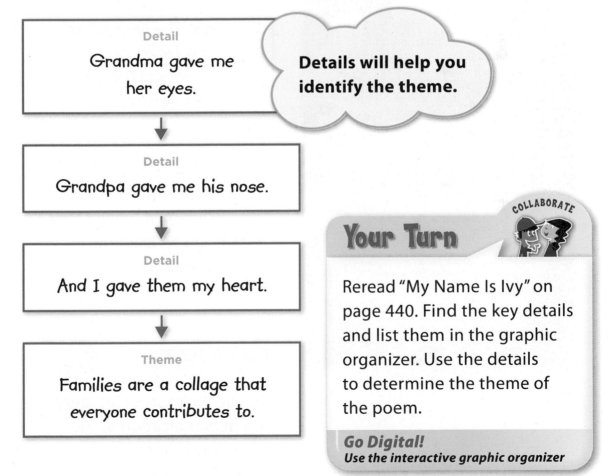

Detail

Grandma gave me her eyes.

Details will help you identify the theme.

↓

Detail

Grandpa gave me his nose.

↓

Detail

And I gave them my heart.

↓

Theme

Families are a collage that everyone contributes to.

Your Turn COLLABORATE

Reread "My Name Is Ivy" on page 440. Find the key details and list them in the graphic organizer. Use the details to determine the theme of the poem.

Go Digital!
Use the interactive graphic organizer

Imagery and Personification

Imagery is the use of specific language to create a picture in a reader's mind. **Personification** is giving human qualities to a non-human thing such as an animal or object.

 Find Text Evidence

I can find examples of imagery and personification when I reread the poem "Climbing Blue Hill" on page 439.

page 439

CLIMBING BLUE HILL

When the yellow leaves begin to
glimmer among the green ones,
we hike up Blue Hill
through an early morning mist.

"It's not much farther, boys!"
My grandfather bellows happily,
his words an echo of all the other times
he's had to urge us up a steep trail.

I hear the comforting squeak of his boots
as the ground's chill breath whispers
against our ankles and the overgrown
branches tug curiously at my hair.

Abruptly, the trail spits us out,
onto gray rock, into blue sky and sunlight.
My brother shouts, shoves me aside,
races to the low bushes huddled against the wind.

His fingers tug at the tiny leaves.
"Look! Blueberries!" He yells.
And we gobble the blue sweetness up,
my brother, my grandfather, and me.

— Andrew Feher

Imagery The lines *When the yellow leaves begin to/ glimmer among the green ones* are an example of imagery.

Personification The lines *as the ground's chill breath whispers/against our ankles* are an example of personification.

COLLABORATE

Your Turn

Find an example of imagery and personification in the poem "My Name Is Ivy."

444

Figurative Language

A **metaphor** is a comparison of two unlike things without the use of *like* or *as*.

 ### Find Text Evidence

To find a metaphor, I need to look for two unlike things that are being compared. In the poem "Collage," on page 441, the grandmother compares the grandfather's nose to a road.

Grandpa gave me his nose.
"A bumpy, rocky road of a nose,"
Grandma scoffs.

 **Your Turn** COLLABORATE

Reread "My Name Is Ivy" on page 440. What is the central metaphor in the poem?

Susan Gal

445

Readers to ...

Writers use descriptive and concrete details to help the readers build a picture in their minds. Reread the first two stanzas of "Climbing Blue Hill."

Supporting Details

Identify the **supporting details**. How do these details help the reader picture what is happening?

Expert Model

When the yellow leaves begin to
glimmer among the green ones,
we hike up Blue Hill
through an early morning mist.

"It's not much farther, boys!"
My grandfather bellows happily,
his words an echo of all the other times
he's had to urge us up a steep trail.

Susan Gal

446

Writers

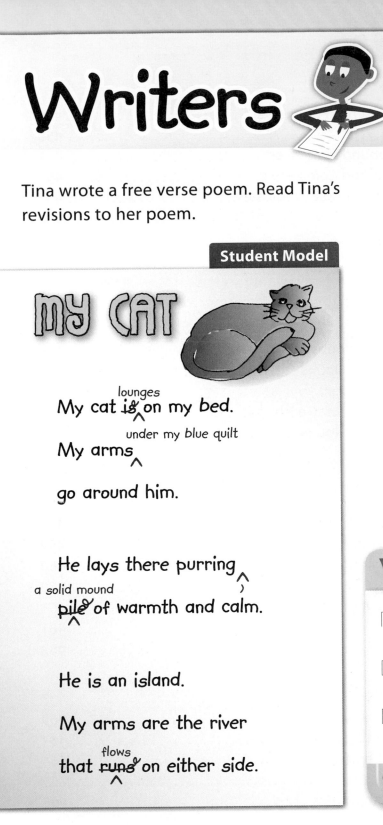

Tina wrote a free verse poem. Read Tina's revisions to her poem.

Student Model

MY CAT

My cat ~~is~~ on my bed.
^{lounges}

My arms ^{under my blue quilt}

go around him.

He lays there purring ‸

a solid mound
~~pile~~ of warmth and calm.

He is an island.

My arms are the river

that ~~runs~~ on either side.
^{flows}

Editing Marks

⌐⌐ Switch order.

∧ Add.

∧ Add a comma.

⟍ Take out.

(SP) Check spelling.

≡ Make a capital letter.

Grammar Handbook

Sentences Using Prepositions
See page 471.

Your Turn

COLLABORATE

✔ Identify the descriptive details Tina added.
✔ Identify the sentences with prepositions.
✔ Tell how the revisions improved her writing.

Go Digital!
Write online in Writer's Workspace

447

Contents

Sentences

Sentences and Sentence Fragments 450
Sentence Types 450
Simple and Compound Sentences 451
Complete Subjects and Complete Predicates 451
Simple Subjects and Simple Predicates 452
Compound Subjects and Compound Predicates 452
Complex Sentences 453
Run-On Sentences...................................... 454

Nouns

Singular and Plural Nouns............................... 455
More Plural Nouns...................................... 455
Common and Proper Nouns.............................. 456
Concrete and Abstract Nouns 456
Collective Nouns 456
Singular and Plural Possessive Nouns 457
Combining Sentences: Nouns 457

Verbs

Action Verbs 458
Verb Tenses 458
Subject-Verb Agreement................................ 459
Spelling Present- and Past-Tense Verbs 459
Main Verbs and Helping Verbs........................... 460
Helping Verbs: *has, have, had*......................... 460
Helping Verbs: *can, may, must* 461
Linking Verbs 461
Irregular Verbs 462

Pronouns

Pronouns 463
Pronoun-Antecedent Agreement 463
Reflexive Pronouns 464
Pronoun-Verb Agreement................................ 464
Possessive Pronouns................................... 465
Pronouns and Homophones 465

Adjectives

Adjectives 466
Articles 466
This, That, These, and *Those* 466
Adjectives That Compare 467
Comparing: *More* and *Most, Good* and *Bad* 467

Adverbs

Adverbs 468
Using *Good* and *Well* 468
Adverbs That Compare 469
Comparing with Irregular Adverbs 469

Negatives

Negatives and Negative Contractions 470
Double Negatives 470

Prepositions

Prepositions 471
Prepositional Phrases 471

Mechanics: Abbreviations

Titles and Names 472
Time 472
Days and Months 473
Addresses 473

Mechanics: Capitalization

First Words in Sentences 474
Letter Greetings and Closings 474
Proper Nouns: Names and Titles of People 475
Titles of Works 475
Other Proper Nouns and Adjectives 476

Mechanics: Punctuation

End Punctuation 477
Periods 477
Colons and Semicolons 478
Apostrophes 478
Parentheses 478
Commas 479
Quotation Marks 480
Italics (Underlining) 480

Sentences

Sentences and Sentence Fragments

A **sentence** expresses a complete thought. A **sentence fragment** does not express a complete thought.

Al writes about the storm. (complete sentence)

The heavy rains. (needs a predicate)

Your Turn Write each group of words. Write sentence or fragment next to it to identify each item. Then rewrite each fragment to make a complete sentence.

1. We listened to the news reports.

2. The strong winds.

Sentence Types

Each of the four types of sentences begins with a **capital letter** and ends with an **end mark**.

A **declarative sentence** makes a statement. It ends with a **period**.	*Scott rode a horse last week.*
An **interrogative sentence** asks a question. It ends with a **question mark**.	*Did you see him on the trail?*
A **imperative sentence** tells or asks someone to do something. It ends with a **period**.	*Take a picture of the group.*
An **exclamatory sentence** shows strong feeling. It ends with an **exclamation mark**.	*We had a great time riding!*

Your Turn Write each sentence. Add the correct punctuation. Then write what kind of sentence it is.

1. I had never been on a horse

2. Do you know how to ride a horse

Simple and Compound Sentences

A **simple sentence** has only one complete thought. A **compound sentence** has two or more complete thoughts. The **coordinating conjunctions** *and, but,* and *or* connect the complete thoughts in a compound sentence.

> *My mother works in the city.* (simple sentence)
> *My mother works in the city, **but** my father works at home.*
> *(compound sentence)*

Your Turn Write each sentence. Then tell whether the sentence is *simple* or *compound*. If it is a compound sentence, circle the coordinating conjunction.

1. My brother volunteers at the library.
2. He works on Saturdays, and Mom drives him there.
3. I would join him, but my team practices that day.
4. You can walk home, or Dad can pick you up.
5. Some days everyone is doing different things.

Complete Subjects and Complete Predicates

Every sentence has two important parts: the **subject** and the **predicate**.

The **subject** tells whom or what the sentence is about. The **complete subject** is all the words in the subject part.

> ***The woman next door*** *works in her garden.*

The **predicate** tells what the subject does or is. The **complete predicate** is all the words in the predicate.

> *The woman next door **works in her garden**.*

Your Turn Write each sentence. Underline the complete subject. Circle the complete predicate.

1. Our neighbor grows his own vegetables.
2. He plants three different gardens.
3. My sister and I help.
4. Will he invite us to dinner?
5. His recipe for tomato soup is the best!

Simple Subjects and Simple Predicates

The **simple subject** is the main word in the complete subject.

*Our **vacation** in Boston starts on Saturday.*

The **simple predicate** is the main word in the complete predicate.

*Our vacation in Boston **starts** on Saturday.*

Your Turn Write each sentence. Underline the simple subject. Circle the simple predicate.

1. The dog charged after the rabbit.

2. The dog's owner then chased after him.

Compound Subjects and Compound Predicates

A **compound subject** contains two or more simple subjects that have the same predicate.

*The **cat** and **dog** ran outside.*

A **compound predicate** contains two more simple predicates that have the same subject.

*The cat **sat** and **meowed** at the door.*

Use the **conjunction** *and* or *or* to combine sentences and create compound subjects or compound predicates. When you combine three or more simple subjects or simple predicates, use **commas** to separate them.

Jan stepped outside. Jan turned left. Jan headed home.
Jan stepped outside, turned left, and headed home.
(compound predicate)

Your Turn Combine the sentence pairs to form one sentence. Then write whether the new sentence has a compound subject or a compound predicate.

1. Dad raked the yard. Mom raked the yard.

2. Mom mowed the lawn. Mom washed the car.

Complex Sentences

A **complex sentence** contains an independent clause and one or more dependent clauses.

We pitched our tent where the ground was flat and dry.

An **independent clause** can stand alone as a sentence.

We pitched our tent.

A **dependent clause** cannot stand alone as a sentence and begins with a **subordinating conjunction**. Some common subordinating conjunctions are *after, although, before, because, during, if, since, until, when, where,* and *while.*

where the ground was flat and dry

Use a **comma** after the dependent clause when it comes at the beginning of a sentence.

After the sun went down, we heard an owl.

Your Turn Write each sentence. Underline the independent clause. Circle the dependent clause.

1. We looked up at the sky after it got dark.
2. You might see a shooting star if you wait long enough.
3. While we were watching, the Moon rose above the trees.
4. When I grow up, I might become an astronaut.
5. We talked until our parents told us to go to sleep.

Run-On Sentences

A **run-on sentence** contains two or more independent clauses without the proper conjunctions or punctuation.

I dropped a book my cat got scared it ran away.

You can correct run-on sentences using one or more strategies.

Break the independent clauses into separate sentences.	*I dropped a book. My cat got scared. It ran away.*
Create a compound subject or compound predicate.	*I dropped a book. My cat got scared and ran away.*
Create a compound sentence using coordinating conjunctions.	*I dropped a book, and my cat got scared and ran away.*
Create a complex sentence using subordinating conjunctions.	*Because I dropped a book, my cat got scared and ran away.*

Your Turn Correct each run-on sentence using one or more of the strategies described above.

1. My alarm didn't go off I missed the bus.
2. Mom was already at work I had to walk.
3. The sun was out it was really chilly.
4. I got to school I raced up to the door.
5. I was so embarrassed it was closed it was Saturday!

Nouns

Singular and Plural Nouns

A **noun** names a person, place, thing, or idea. It can be a single word or a group of words used together. A **singular noun** names one person, place, thing, or idea. A **plural noun** names more than one. Add -*s* to form the plural of most nouns. Add -*es* to form the plural of nouns ending in *s*, *x*, *ch*, or *sh*.

Singular nouns: boy   school   home run   joy
Plural nouns: girls   stores   churches   beliefs

Your Turn **Write each sentence. Underline each noun and write whether it is singular or plural.**

1. The actor needed to learn his lines.
2. His friend read from the script.

More Plural Nouns

If a noun ends in a consonant + *y*, change *y* to *i* and add -*es*.	*ladies, berries, skies, libraries*
If a noun ends in a vowel + *y*, add -*s*.	*boys, monkeys, days, essays*
If a noun ends in -*f*, you may need to change *f* to *v* and add -*es*.	*chefs, roofs, leaves, hooves, knives*
Some nouns have the same singular and plural forms.	*deer, sheep, moose, fish*
Some nouns have special plural forms.	*men, women, children, teeth, feet*

Your Turn **Write each sentence. Change the singular noun in parentheses ( ) into a plural noun.**

1. The park is crowded on (holiday).
2. Many (child) visit with their parents.
3. If there are sunny (sky), they play outside.

455

Common and Proper Nouns

A **common noun** names any person, place, or thing. A **proper noun** names a particular person, place, or thing. A proper noun always begins with a capital letter.

> The **student** looked at the **map**. (common)
> **Brittany** located **Ohio**. (proper)

Your Turn Write each sentence. Underline each noun and write whether it is common or proper.

1. Anna has a map of the world on her wall.
2. There are pins placed in several countries.

Concrete and Abstract Nouns

A **concrete noun** names a person, place, or thing that physically exists and can be perceived with the senses. An **abstract noun** names a quality, concept, or idea that does not physically exist. Many abstract nouns have no plural form.

> **Ellen** set the **sheet music** on the **piano**. (concrete)
> **Music** fills the **soul** with **happiness**. (abstract)

Collective Nouns

A **collective noun** names a group acting as a single unit. Collective nouns can also have plural forms.

> Our **team** plays three other **teams** next week.

Your Turn Write each sentence. Underline each noun and write whether it is concrete or abstract. Circle any collective nouns.

1. The road crew stopped traffic on our street.
2. My sisters wanted to go to the shopping mall.
3. Their car was stopped for a long time.

Singular and Plural Possessive Nouns

A **possessive noun** is a noun that shows who or what owns or has something.

A singular possessive noun is formed by adding an **apostrophe (')** + s to the end of a singular noun.

The car's alarm made the boy's ears hurt.

Most plural nouns ending in -s become possessive by adding an apostrophe to the end. Irregular plural nouns that don't end in -s add an apostrophe + s.

The visitors' center hosted a children's reception.

Your Turn **Write each sentence. Change the word in parentheses into a possessive noun.**

1. My (brother) band practices in the garage.
2. The (group) poster shows their lead singers.
3. The two (singers) voices sound alike.
4. They played at our (school) fall festival.
5. All of the (newspapers) reviews were good.

Combining Sentences: Nouns

You can combine nouns in the subject.

Sean went inside. Cora went inside.
Sean and Cora went inside.

You can combine nouns in the predicate.

Mom likes checkers. Mom likes chess.
Mom likes checkers and chess.

Your Turn **Combine the nouns in the sentence pairs to form one sentence.**

1. Chet went to the shore. Emily went to the shore.
2. They saw sea gulls. They saw pelicans.
3. A boy swam nearby. An older woman swam nearby.
4. The snack bar sold juice. The snack bar sold fruit.
5. Were your friends there? Were your parents there?

Verbs

Action Verbs

An **action verb** is a word that expresses action. It tells what the subject does or did.

*The pitcher **threw** the ball over the plate.*

Your Turn **Write each sentence. Underline the action verb(s).**

1. Volunteers gathered in the city park.
2. They painted the information booth.
3. One crew repaired all the picnic tables.

Verb Tenses

A **present-tense verb** shows action that happens now.
*Today, the lifeguard **watches** the weather.*

A **past-tense verb** shows action that has already happened.
*Yesterday, the lifeguard **listened** for thunder.*

A **future-tense verb** shows action that may or will happen.
*At the first sign of lightning, he **will close** the pool.*

A **progressive tense** shows action that continues over time. Use the verb *be* with the *-ing* form of another verb to create the **present progressive, past progressive,** or **future progressive** tense.

*I **am watching** you. (present progressive)*
*I **was watching** you. (past progressive)*
*I **will be watching** you. (future progressive)*

Your Turn **Write each sentence. Underline the verb and tell what tense it is.**

1. Next summer I will attend space camp.
2. I received a brochure in the mail last week.
3. My mother helps me with the registration process.
4. I will be checking the camp's Web site every day now.
5. I am counting the days until the start of the program.

Subject-Verb Agreement

A present-tense verb must agree with the subject of the sentence. Add -s to most verbs if the subject is singular. Add -es to verbs that end in s, ch, sh, x, or z. Do not add -s or -es if the subject is plural or I or you.

*Cristen **watches** the race. Her friends **cheer** for her sister.*

Your Turn Write each sentence. Use the correct form of the verb(s) in parentheses.

1. The racers (crouch) at the starting line.
2. The clock (count) down the seconds.
3. Cristen's sister (rush) off to a great start.
4. She (take) the lead right away.
5. Another runner (catch) up and (challenge) her.

Spelling Present- and Past-Tense Verbs

The spelling of some verbs changes when -es or -ed is added. For verbs ending in a consonant + y, change the y to i before adding -es or -ed. For verbs ending in one vowel and one consonant, double the final consonant before adding -ed. For verbs ending in e, drop the e before adding -ed.

*Josh **cried** when the music **stopped**. He truly **loved** that song.*

Your Turn Write each sentence. Use the correct form of the verb in parentheses.

1. Now Josh (carry) his guitar upstairs.
2. He (try) to remember the song.
3. Yesterday he (practice) all day.
4. He (step) up to the challenge and worked hard.
5. Who (worry) that he may forget the words?

Main Verbs and Helping Verbs

The **main verb** in a sentence tells what the subject does or is. A **helping verb** helps the main verb show an action or make a statement. The verb *be* is often used as a helping verb.

Carol **is running** for class president.

Use a **contraction** to combine a helping verb with the subject or with the word *not*. An **apostrophe (')** takes the place of the missing letters.

Carol's feeling sad. Her friends **aren't** helping much.

Your Turn Write each sentence. Underline the main verbs and circle the helping verbs. If there is a contraction, tell which two words have been combined.

1. Many students are making posters.
2. Carol is preparing a speech.
3. Her school is holding a debate tomorrow.
4. Reporters were writing about it last week.
5. Carol's hoping that she will win the election.

Helping Verbs: *has, have, had*

The helping verbs **has, have,** and **had** can be used with the past-tense form of a verb to show an action that has already happened. When *has* or *have* is used, the entire verb forms the **present perfect tense**. When *had* is used, the entire verb forms the **past perfect tense**.

She **has eaten** all of the bread he **had baked**.

Your Turn Write each sentence. Choose the correct form of **has, have,** or **had** to complete the sentence.

1. He (have, had) cooked every day last week.
2. Today she (has, had) watched him make dried fruit.
3. (Have, Has) you ever tasted anything so delicious?

Helping Verbs: *can, may, must*

The verbs *can, may,* and *must* can be used as **helping verbs** with a main verb.

> They **may wait** until someone **can help** them.

Your Turn Write each sentence. Underline each main verb and circle each helping verb.

1. We may buy a gift for our parents.
2. What can we afford?
3. There must be something good here!
4. "Can I borrow some money?" I ask.
5. You reply, "We must stay within our budget."

Linking Verbs

A **linking verb** links the subject to a noun or adjective in the predicate. It must agree with the subject. A linking verb does not express action. Some common present-tense linking verbs are *am, are,* and *is.* Some common past-tense linking verbs are *was* and *were.*

> Carlos **is** an artist. His exhibits last year **were** beautiful.

Your Turn Write each sentence. Underline each linking verb.

1. This painting is my favorite.
2. The colors are bright and joyful.
3. Carlos is proud of his most recent work.
4. I am anxious to see it.
5. Was he happy with the review on the Web?

Irregular Verbs

An **irregular verb** is a verb that does not end in -ed to form the past tense. Some also have special spellings when used with the helping verb *have*.

Present	Past	With *Have*
begin	began	begun
bring	brought	brought
come	came	come
do	did	done
draw	drew	drawn
eat	ate	eaten
give	gave	given
go	went	gone
grow	grew	grown
hide	hid	hidden
run	ran	run
say	said	said
see	saw	seen
sing	sang	sung
sit	sat	sat
take	took	taken
tell	told	told
think	thought	thought
write	wrote	written

Your Turn Write each sentence. Use the correct form of the verb in parentheses.

1. Last night I (go) to a spelling bee.
2. I (tell) my friend to meet me there.
3. We have (see) several spelling bees together.

Pronouns

Pronouns

A **pronoun** is a word that takes the place of one or more nouns.
A **subject pronoun** is used as the subject of a verb. It tells who or what does the action. The pronouns *I, you, he, she, it, we,* and *they* can be used as subject pronouns.

An **object pronoun** is used as the object of a verb. It tells whom or what received the action of the verb. The pronouns *me, you, him, her, it, us,* and *them* can be used as object pronouns.

An object pronoun may come after prepositions such as *for, at, of, with,* or *to.*

> *I gave **him** the hammer. **He** used **it** to build a shelf for **her.***

Your Turn Write each sentence. Underline each subject pronoun. Circle each object pronoun.

1. She took the little statues out of the boxes.
2. Then she placed them on the new shelves.
3. They looked much better to her.
4. We went over to see what she had done.
5. You and he should invite her to dinner.

Pronoun-Antecedent Agreement

A **pronoun** must match its **antecedent**, the noun to which it refers. The antecedent may or may not be in the same sentence.

> ***Mom** said **she** knew what to do. **She** called my **brother** and gave **him** advice.*

Your Turn Write each sentence. Underline each pronoun. Circle each antecedent.

1. My brother asked Mom to drive him to work.
2. Dad printed the directions and gave them to Mom.
3. Dad was worried about my brother. He asked to go along.
4. Dad and Mom told my brother to expect them soon.

Reflexive Pronouns

A **reflexive pronoun** tells about an action that a subject does for or to itself. A reflexive pronoun is based on an object pronoun because it receives the action of the verb. The ending **-self** is added for singular pronouns. The ending **-selves** is added for plural pronouns.

*The girl wrote **herself** a note. We drove **ourselves** to the city.*

Your Turn Write each sentence. Underline only the reflexive pronoun(s).

1. We got ourselves lost in the city.
2. I asked myself how it happened.
3. You never expect to find yourself in trouble.
4. Trouble can find you all by itself.
5. My sister cheered for herself when she found a map.

Pronoun-Verb Agreement

A present-tense verb must **agree** with its subject, even if the subject is a pronoun.

*I **am** thirsty. She **is** thirsty. We **are** thirsty.*

Your Turn Write each sentence. Use the correct present-tense form of the verb in parentheses.

1. He (look) for a water fountain.
2. She (find) one over by the tennis courts.
3. We (proceed) from there to the parking lot.
4. Our parents beep the horn when they (see) us.
5. She (climb) into the front seat and (say), "Let's go!"

Possessive Pronouns

A **possessive pronoun** takes the place of a possessive noun. It shows who or what owns something. *My, your, her, his, its, our,* and *their* are possessive pronouns.

I gave **my** *order to* **our** *waiter. He wrote it on* **his** *pad.*

Some possessive pronouns can stand on their own. *Mine, yours, hers, his, its, ours,* and *theirs* can be used alone.

We order lunch. I won't eat **mine** *until* **yours** *is here.*

Your Turn **Write each sentence. Replace the words in parentheses with a possessive pronoun.**

1. (My friend Lauren's) sandwich looked very tasty.
2. Hers had more peppers than (the sandwich belonging to me).
3. "Can I have a bite of (the sandwich belonging to you)?" she asked.

Pronouns and Homophones

Some possessive pronouns sound like pronoun-verb contractions but are spelled differently. A possessive pronoun does not contain an apostrophe because no letters are missing.

It's *time for the class to take* **its** *test.* (**It's** = It + is; **its** is a possessive pronoun)

You're *proud of* **your** *grade.* (**You're** = You + are; **your** is a possessive pronoun)

They're *not pleased with* **their** *grades.* (**They're** = They + are; **their** is a possessive pronoun)

There's *no excuse for* **theirs***.* (**There's** = There + is; **theirs** is a possessive pronoun)

Your Turn **Write each sentence. Choose the correct word in parentheses to complete the sentence.**

1. "You're not going to believe (you're, your) eyes," I said.
2. "(Theirs, There's) a B on my report card."
3. "Is it moving (it's, its) wings?" you asked.

465

Adjectives

Adjectives

An **adjective** describes a noun. Adjectives can tell **what kind** or **how many**. Most adjectives come directly before the nouns they describe. When an adjective comes after the noun it describes, the noun and adjective are connected by a **linking verb**. A **proper adjective** is formed from a proper noun.

My **two favorite** players on the team are **Brazilian**.

Your Turn Write each sentence. Circle each adjective and underline the noun being described.

1. We can't play on the wet fields.
2. The grass is slippery.

Articles

The words **the, a,** and **an** are special adjectives called **articles**. Use a before words that begin with consonant sounds. Use an before words that begin with vowel sounds.

A squirrel dropped the acorns from an oak tree.

Your Turn Write the sentence. Circle each article.

1. The acorns clattered on the tin roof of an old garage.

This, That, These and Those

This, that, these and **those** are special adjectives that tell **how many** and **how near or far away** something or someone is. This (near) and that (far) are used with singular nouns. These (near) and those (far) are used with plural nouns.

These apples in my hand came from **that** tree by the fence.

Your Turn Write the sentence. Choose the correct word in parentheses to complete the sentence.

1. We picked (that, these) apples from (this, those) tree.

Adjectives That Compare

Add **-er** to an adjective to compare two nouns. Add **-est** to compare more than two nouns.

If an adjective ends in a consonant and y, then change the y to i before adding -er or -est.

If an adjective ends in e, then drop the e before adding -er or -est.

If an adjective has a single vowel before a final consonant, then double the final consonant before adding -er or -est.

*I take the **lighter, tinier** box. The other box is the **thinnest** of all.*

Your Turn Write each sentence. Use the correct form of the adjective in parentheses.

1. My room is (big) than my sister's room.

2. "It's the (messy) room in the house!" Mom said.

Comparing: *More* and *Most, Good* and *Bad*

Use **more, better**, and **worse** to compare two people, places, or things.

Use **most, best,** and **worst** to compare more than two people, places, or things.

*I have **many** coins. You have **more** coins than I do. He has the **most** coins of all.*

*Eli is a **good** chef. Pam is a **better** chef than Eli. Chad is the **best** chef of all.*

*She had a **bad** day. He had a **worse** day than she did. I had the **worst** day of all.*

Use **more** and **most** with longer adjectives instead of adding the endings -er and -est.

*Mom had a **more pleasant** trip than I did. Dad had the **most exciting** trip of all.*

Your Turn Write each sentence. Use the correct form of the adjective in parentheses.

1. This restaurant has the (good) soup in town.

2. There were (many) choices today than yesterday.

Adverbs

Adverbs

An **adverb** is a word that tells more about a verb. Adverbs often tell *how, when,* or *where.* Many adverbs end in *-ly.*

> You **nervously** looked **up**. The storm would start **soon**.

Your Turn Write each sentence. Circle each adverb and draw a line under the verb that each adverb describes.

1. Clouds gathered overhead.
2. Lightning flashed and thunder rumbled loudly.
3. We quickly raced inside.
4. "Is everyone here?" our father asked.
5. You quietly hoped the storm would go away.

Using *Good* and *Well*

Good is an adjective that tells more about a noun. **Well** is an adverb that is tells more about a verb.

> The **good** nurse treated us **well**.

Your Turn Write each sentence. Choose *good* or *well* to complete the sentence. Underline the word that is being described.

1. There is a (good, well) chance that we will miss the play.
2. My mother did not sleep (good, well).
3. When I am sick, I can't perform (good, well).
4. Do you have any (good, well) suggestions?
5. A (good, well) actor might pretend to be healthy.

Adverbs That Compare

Adverbs can be used to compare two or more actions. Use *more* before most adverbs to compare two actions. Use *most* before most adverbs to compare more than two actions. Add the ending -*er* or -*est* to shorter adverbs to compare actions.

> Gil danced **longer** and **more gracefully** than Wendy.
> Shawna danced **longest** and **most gracefully** of all.

Your Turn **Write each sentence. Choose the correct form of the adverb in parentheses.**

1. Gil can move (swiftly) than Wendy.
2. Shawna works (hard) than Wendy.
3. Wendy learns new dances (fast) of all.
4. Have you danced (recently) than Gil?
5. Shawna has performed (consistently) of all the dancers.

Comparing with Irregular Adverbs

With the adverb **well**, use **better** to compare two actions. Use **best** to compare more than two actions.
With the adverb **badly**, use **worse** to compare two actions. Use **worst** to compare more than two actions.

> *He and I played worse than yesterday, but you played best of all.*

Your Turn **Write each sentence. Choose the correct form of the adverb in parentheses.**

1. I ran (well) at this race than the last one.
2. You had prepared (well) of all.
3. He reacted (badly) to the score than I did.
4. Their coach behaved (badly) of all.
5. Our coach understands the game (well) than any other coach.

Negatives

Negatives and Negative Contractions

A **negative** is a word that means *no*. Many negatives contain the word *no* within them. Some negatives use the contraction *n't*, which is short for *not*.

> **Nobody** wants to go first. I **can't** understand why.

Your Turn Write each sentence. Underline the negative word in each one.

1. We could not see anything inside the room.
2. There was no light switch on the wall.
3. I don't like going into a dark room.
4. I never enter a room that isn't lit brightly.
5. The lamps were nowhere to be found.

Double Negatives

Do not use two negatives in one sentence.

> Don't you (**ever**, never) talk to me like that!
> I don't think (no one, **anyone**) should behave that way.

Your Turn Write each sentence. Choose the correct word in parentheses to complete the sentence.

1. No one has (never, ever) won an argument with my mother.
2. We didn't have (no, any) idea what to say to her.
3. She won't take (any, none) of our advice.
4. Nothing (will, won't) make her change her mind.
5. Won't she trust (no one, anyone) other than herself?

Prepositions

Prepositions

A **preposition** comes before a noun or a pronoun. A preposition shows how the noun or pronoun is linked to another word in the sentence. Some common prepositions are *in, at, of, from, with, to,* and *by.*

> The conductor **on** the train waved **at** the boy.

Your Turn Write each sentence. Circle each preposition.

1. He gave his ticket to the conductor.
2. The train left from the station at noon.
3. He sat by the window with his mother.
4. The motion of the train shook his belongings.
5. The book with the blue cover fell from his backpack.

Prepositional Phrases

A **prepositional phrase** is a group of words that begins with a preposition and ends with a noun or pronoun. The noun or pronoun is the **object of the preposition**. A prepositional phrase can be used as an adjective or an adverb in a sentence.

> The girl **in the park** (adjective) hit the ball **over the net** (adverb).

Your Turn Write each sentence. Underline each prepositional phrase and circle each preposition. Then place an "O" above the object of the preposition.

1. The ball bounced at her feet.
2. The girl with the long hair kept score.
3. They played for three hours.
4. My friend and I cheered from the bleachers.
5. My favorite one of the players scored the last point.

471

Mechanics: Abbreviations

Titles and Names

Some titles are **abbreviations**, or shortened forms of words. Other titles, like *Ms.* and *Mrs.*, don't have longer forms. An **initial** is the first letter of a name. Titles and initials are capitalized and are followed by a period. When abbreviations are used at the end of an Internet address, they are not capitalized or followed by a period.

> *Dr. A. J. Moreno will post Sen. Paulsen's speech on our state's .gov Web site.*

Your Turn **Write each sentence. Change the word(s) in parentheses into an abbreviation or initial.**

1. I sent an e-mail to (Mister) Elish.
2. (Governor) Slater also wrote a response.
3. Ms. (Carol Jane) Stein will speak to our class next week.
4. I posted the news at www.ourschool.(educational) today.
5. Let's give a warm welcome to (Doctor) and (his wife) Yee.

Time

Use abbreviations to indicate time before noon (A.M. for "ante meridiem") and after noon (P.M. for "post meridiem"). These abbreviations are capitalized with periods after each letter.

> *Our car wash will go from 10 **A.M.** to 2 **P.M.** on Saturday.*

Your Turn **Write each sentence. Use the correct abbreviation to replace the words in parentheses.**

1. We will take a break for lunch at 12:30 (after noon).
2. I don't have to be there until 11:00 (before noon).
3. The dog usually waits until 7:30 (in the morning) to wake me up.
4. Mom will pick us up at 2:30 (in the afternoon).
5. I'll be able to walk the dog at 8:00 (in the evening).

Days and Months

When you abbreviate the **days of the week** or the **months of the year**, begin with a capital letter and end with a period. Do not abbreviate *May, June,* or *July.*

Sun.   Mon.   Tues.   Wed.   Thurs.   Fri.   Sat.
Jan.   Feb.   Mar.   Apr.   Aug.   Sept.   Oct.   Nov.   Dec.

Your Turn **Write each sentence. Use the correct abbreviation(s) to replace the word(s) in parentheses.**

1. I have a piano lesson each (Tuesday) in July.
2. Soccer practice only goes until (November) this year.
3. The drama club uses the theater each (Monday) in (March) for rehearsals.
4. The final production begins in (April) and runs until (June).
5. I've blocked out (Wednesday) through (Friday) for vacation.

Addresses

Address abbreviations are capitalized and followed by a period. Some common address abbreviations are **St.** (Street), **Rd.** (Road), **Ave.** (Avenue), **Dr.** (Drive), **Blvd.** (Boulevard), **Ln.** (Lane), **Apt.** (Apartment), and **P. O.** (Post Office), When you write an address, you may use United States Postal Service abbreviations for the names of states. All of these abbreviations are two capital letters with no period at the end. When using these state abbreviations, no comma is needed after the name of the city or town.

She mailed the postcard to 28 Irving **Dr.,** **Apt.** *4B, Canton* **OH.**

Your Turn **Write each address. Use the correct abbreviation(s) whenever possible.**

1. 6 Main Street
2. Post Office Box 1023
3. Providence, Rhode Island
4. 532 Jefferson Street, Los Angeles, California
5. 104 7th Avenue, Apartment 8C, New York, New York

Mechanics: Capitalization

First Words in Sentences

Capitalize the first word of a sentence. Capitalize the first word of a direct quotation. Do not capitalize the second part of an interrupted quotation. When the second part of a quotation is a new sentence, put a period after the interrupting expression and capitalize the first word of the new sentence.

"Finish your homework," my mother said, "and come down for supper."

Dinner smelled great. "I'll be right there," I replied. "I'm almost done."

Your Turn **Write each sentence. Use capital letters correctly.**

1. the final problem was taking a long time to answer.
2. "did you hear what I said?" asked my mother.
3. without looking up, I replied, "yes, I did."
4. "don't wait too long," Mom said, "or it will get cold."
5. "this is too hard," I said. "maybe I can finish it later."

Letter Greetings and Closings

All of the words in a letter's greeting begin with a capital letter. Only the first word in the closing of a letter begins with a capital letter.

Dear Dr. Watkins, Sincerely yours,

Your Turn **Write each part of a letter with the correct capitalization.**

1. dear uncle floyd,
2. best wishes
3. with all my love,
4. dear ladies and gentlemen,
5. to whom it may concern:

Proper Nouns: Names and Titles of People

Capitalize the names of people and the initials that stand for their names. Capitalize titles or abbreviations of titles when they come before or after the names of people. Capitalize words that show family relationships when used as titles or as substitutes for a person's name. Do not capitalize words that show family relationships when they are preceded by a possessive noun or pronoun. Capitalize the pronoun *I*.

> *Dean's father and I talked to Mom about his visit to*
> *Dr. T. J. Hunter, Jr.*

Your Turn **Write each sentence. Use capital letters correctly.**

1. mr. weston worried about Dean's swollen knee.
2. "I'll ask dad what to do about it," Dean had told me.
3. "What did dr. hunter tell him?" i asked my mother.
4. "He was sent to another doctor," mom explained.

Titles of Works

Capitalize the first, last, and all important words in the title of a book, play, short story, poem, movie, article, newspaper, magazine, TV series, chapter of a book, or song.

> *My father sang "Moon River" while I watched "Alice in*
> *Wonderland" again.*
> *The book <u>Give Us a Chance</u> was reviewed in today's*
> *<u>Tarrytown Tribune</u>.*

Your Turn **Write each sentence. Use capital letters correctly.**

1. We sang "america the beautiful" at the start of the game.
2. A reporter from <u>the hometown herald</u> wrote about it.
3. He compared the victory to the movie "the miracle team."
4. I wrote a poem about it called "winning by the book."

Other Proper Nouns and Adjectives

Capitalize the names of cities, states, countries, and continents. Do not capitalize articles or prepositions that are part of the names. Capitalize the names of bodies of water and geographical features. Capitalize the names of sections of the country. Do not capitalize compass points when they just show direction.

Portland, Oregon   *California is south of the Pacific Northwest.*

Capitalize the names of streets and highways. Capitalize the names of buildings, bridges, and monuments.

Mackinaw Bridge   *Empire State Building*

Capitalize the names of stars and planets. Capitalize *Earth* when it refers to the planet. Do not capitalize *earth* when it is preceded by the article *the*. Do not capitalize *sun* or *moon*.

The planet next closest to the sun from the earth is Venus.

Capitalize the names of schools, clubs, teams, businesses, and products.

Junior Debate Club at Westwood Senior High School

Capitalize the names of historic events, periods of time, and documents.

the Battle of Bunker Hill   *the Declaration of Independence*

Capitalize the days of the week, months of the year, and holidays.

Labor Day is the first Monday in September.

Capitalize the names of ethnic groups, nationalities, and languages. Capitalize proper adjectives that are formed from the names of ethnic groups and nationalities.

The official languages of the Swiss include German and French.

Capitalize the first word of each main topic and subtopic in an outline.

1. Products and exports
   A. Natural resources

Mechanics: Punctuation

End Punctuation

A **declarative sentence** makes a statement. It ends with a **period (.)**.

An **interrogative sentence** asks a question. It ends with a **question mark (?)**.

An **imperative sentence** makes a command or a request. It ends with a **period (.)** or an **exclamation mark (!)**.

An **exclamatory sentence** expresses strong emotion. It ends with an **exclamation mark (!)**.

> *Do you like trapeze artists? Watch how daring they are! I'm afraid of heights like that.*

Your Turn **Write each sentence. Use the correct capitalization and end punctuation.**

1. When is the next performance
2. Get tickets now before they sell out
3. I can't wait to see the fire-breathing acrobats
4. Her plane flies east from chicago on valentine's day.

Periods

Use a period at the end of an abbreviation. Use a period in abbreviations for time. Use a period after initials. Use a period after numbers and letters in an outline.

> *Dr. E. J. Simmons will see us at 4:45 P.M. on Feb. 23.*

Your Turn **Write each sentence. Insert periods where needed.**

1. I would prefer an appointment at 10:00 AM.
2. Ms Etchells has scheduled the test for Oct 5.
3. My best friend, B D Shea, will park at Elm St and wait.
4. Is 7:30 PM too late for Dr West to see you?

Colons and Semicolons

Use a **colon** to separate the hour and minute when you write the time of day. Use a colon after the greeting of a business letter. Use a **semicolon** to combine two related independent clauses that are not connected by a conjunction such as *or, and,* or *but.*

Dear Professor Cooper:
I cannot make your 10:30 class today; our cat is sick.

Your Turn Write each sentence. Insert the proper punctuation where it is needed.

1. Dear Mr. Kirov
2. I tried to call you this morning no one answered the phone.

Apostrophes

Use an **apostrophe (')** and an *s* to form the possessive of a singular noun. Use an apostrophe and an *s* to form the possessive of a plural noun that does not end in *s.* Use an apostrophe alone to form the possessive of a plural noun that ends in *s.* Do not use an apostrophe in a possessive pronoun. Use an apostrophe in a **contraction** to show where a letter or letters are missing.

*My **friend's** family **didn't** borrow a car because **theirs** was fixed.*

Parentheses

Use parentheses to set off information that is not essential in a sentence, such as unnecessary details, clarifications, or examples.

Jim (winner of last year's contest) didn't register for this year.

Your Turn Write each sentence. Insert apostrophes and parentheses where needed.

1. This years entry fee $25 last year shouldnt increase.
2. Our familys car over 10 years old is very reliable.

Commas

Use a **comma (,)** between the name of a city and the complete name of a state. Use a comma after the name of a state or a country when it is used with the name of a city in a sentence. Do not use a comma between the name of a city and the postal service abbreviation for a state.

We drove from Houston, Texas, to Toronto, Canada, in one week.

Use a **comma** between the day and the year in a date. Use a comma before and after the year when it is used with both the month and the day in a sentence. Do not use a comma if only the month and the year are given.

We gathered on August 5, 1987, for our last reunion.

Use a **comma** after the greeting in a friendly letter and after the closing in all letters.

Dear Aunt Jo,   Very truly yours,

Use a **comma** before *and, but,* or *or* when it joins simple sentences to form a compound sentence. Use a comma to separate two or more subjects in a compound subject. Use a comma to separate two or more predicates in a compound predicate and after a dependent clause at the start of a sentence.

After the bell rang, Liz, Jack, and Chris left, but I remained.

Use a **comma** to set off a direct quotation.

"When you heard the bell," she asked, "had you finished the test?"

Use **commas** to separate three or more items in a series.

She photographed the roses, lilies, and orchids on display.

Use a **comma** after the words *yes* or *no* or other introductory words at the beginning of a sentence. Use a comma with nouns in a direct address.

Yes, I know Brady. By the way, Donna, have you talked to him?

Your Turn Write each sentence. Add commas where needed.

1. The residents of Smith South Carolina wanted a town flag.
2. I submitted six drawings but none of them were chosen.

479

Quotation Marks

Use **quotation marks** before and after the exact words that a speaker says or writes. Use a **comma** or **commas** to separate a clause, such as *he said*, from the quotation itself. Place the comma outside the opening quotation marks but inside the closing quotation marks. Place a **period** inside closing quotation marks. Place a **question mark** or **exclamation mark** inside the quotation marks when it is part of the quotation.

"Did you finish your assignment?" my mother asked.
"I started it," I replied, "but my baby brother interrupted me."

Use **quotation marks** around the title of a short story, song, short poem, print or online article, or chapter of a book.

I wrote a poem called "My Bratty Baby Brother."

Your Turn Write each sentence. Add quotation marks where needed.

1. Our school journal published My Bratty Baby Brother.
2. What gave you the idea for that poem? my teacher asked.
3. It can be so frustrating at home sometimes! I said.

Italics (Underlining)

Use italics or underlining for the title of a book, movie, television series, play, stage production, magazine, or newspaper.

We had tickets to see <u>The Lion King</u> in August.

Your Turn Write each sentence. Underline titles where needed.

1. I borrowed The Big Book of Ballet from the library.
2. We had just watched The Company on television.